AGGRESSION AND WAR
their biological and social bases

AGGRESSION AND WAR
their biological and social bases

EDITED BY

JO GROEBEL

Akademischer Oberrat, Seminar Kommunikationspsychologie, Erziehungswissenschaftliche Hochschule Rheinland-Pfalz, Landau

ROBERT A. HINDE

Royal Society Research Professor and Honorary Director, MRC Unit on the Development and Integration of Behaviour, Madingley, Cambridge

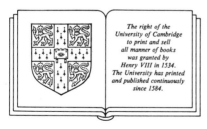

The right of the
University of Cambridge
to print and sell
all manner of books
was granted by
Henry VIII in 1534.
The University has printed
and published continuously
since 1584.

CAMBRIDGE UNIVERSITY PRESS

Cambridge

New York   New Rochelle   Melbourne   Sydney

Published by the Press Syndicate of the University of Cambridge
The Pitt Building, Trumpington Street, Cambridge CB2 1RP
32 East 57th Street, New York, NY 10022, USA
10 Stamford Road, Oakleigh, Melbourne 3166, Australia

First published 1989

Printed in Great Britain at the University Press, Cambridge

British Library cataloguing in publication data

Aggression and war: their biological and social bases
1. Man. Aggression – Sociological perspectives
I. Groebel, J.
II. Hinde, Robert A. (Robert Aubrey), 1923–
302.5′4

Library of Congress cataloguing in publication data

Aggression and war.
Includes index.
1. Aggressiveness (Psychology)
2. Aggressiveness (Psychology) – Physiological aspects.
3. Aggressiveness (Psychology) – Social aspects.
4. War – Psychological aspects.
I. Groebel, Jo.
II. Hinde, Robert A.
[DLMN: 1. Aggression. 2. War, BF 575.A3 A2653]
BF575.A3A513 1989   302.5′4   88-20426

ISBN 0521 35356 4 hard covers
ISBN 0521 35871 X paperback

CE

CONTENTS

CONTRIBUTORS

P. Bateson
Sub-Department of Animal Behaviour
Cambridge University
Madingley, Cambridge CB3 8AA, UK

L. Berkowitz
Department of Psychology
University of Wisconsin
Madison, Wisconsin 53706, USA

N. D. Feshbach
Graduate School of Education
University of California
Los Angeles, California 90024, USA

S. Feshbach
Department of Psychology
1283 Franz Hall
University of California
Los Angeles, California 90024, USA

S. Genovés
Instituto de Investigaciones Antropológicas
Universidad Nacional Autonome de Mexico
04510 Mexico, D.F.

A. P. Goldstein
School of Education
Division of Special

Education and Rehabilitation
Syracuse University
805 South Crouse Avenue
Syracuse, New York 13244, USA

J. H. Goldstein
Department of Psychology
Temple University
Philadelphia, Pennsylvania 19122, USA

J. Groebel
Post-Graduate Program in Communication Psychology
Erziehungswissenschaftliche Hochschule Rheinland-Pfalz
6740 Landau, West Germany

J. Herbert
Department of Anatomy
Cambridge University
Downing Street
Cambridge CB2 3DY, UK

R. A. Hinde
MRC Unit for the Development
    and Integration of Behaviour
Madingley, Cambridge CB3 8AA, UK

F. A. Huntingford
Department of Zoology
University of Glasgow
Glasgow G1 8QQ, UK

K. Lagerspetz
Department of Psychology
University of Turku
Turku, Finland

A. Manning
Department of Zoology
University of Edinburgh
West Mains Road
Edinburgh EH9 3JT, UK

J. M. Rabbie
Faculteit der Sociale Wetenschappen
Rijksuniversiteit te Utrecht
Postbus 80140
3508 TC Utrecht
Heidelberglaan 1, Netherlands

M. H. Segall
Department of Social and
    Political Psychology
Syracuse University
108 Maxwell
Syracuse, New York, 13210, USA

J. D. Singer
Department of Political Science
University of Michigan
Ann Arbor, Michigan 48109, USA

J. M. Winter
Pembroke College
Cambridge CB3 9JH, UK

PREFACE

This book reviews current knowledge on aggression and war at levels of complexity ranging from the physiological, through that of individual aggression, to group conflict and international war. It is especially concerned with two beliefs which, though erroneous, are commonly held in our society. The first is that, because aggressive behaviour is in some sense part of human nature, humans inevitably behave aggressively. The second is that international war is closely related to, and even a direct consequence of, the aggressiveness of individuals. These beliefs both inculcate an unnecessary pessimism about the human condition, and provide an inadequate basis for improving it. They suggest violence can be reduced only by providing harmless outlets for aggressive propensities, or by segregating offenders, and distract us from attempting to build a world where aggressive behaviour is less likely. They present a simplified picture of the bases of international war which conceals its underlying complexity, creates a feeling of individual impotence, and obscures the steps that must be taken to eliminate it. Present explicitly or by implication in political speeches and media reports, these beliefs make a substantial contribution to the maintenance of conditions conducive to violence between individuals, groups and nations.

Disturbed by these issues over a number of years, Santiago Genovés, of the University of Mexico, conceived the idea of issuing a Statement on Violence, comparable with the UNESCO Statement on Race to which he had previously contributed. Beginning in 1980, he and David Adams of the International Society for Research on Aggression began to implement the idea. Their efforts culminated in a meeting of relevant natural and social scientists organized by Martin Ramirez in Seville in May 1986.

Discussion led to the drafting of a Statement, signed by 20 scientists from 12 different countries. Coordinated by David Adams, efforts have been made to disseminate this Statement as widely as possible, and it has already been adopted by a considerable number of professional organizations.

This Statement, reproduced below, was necessarily brief. The aim of this book is to summarize, in non-technical terms, the main part of the evidence on which it was based. At the same time it attempts to provide the interested reader with an overview of empirical research on aggression and war. To this end, experts from diverse scientific fields were asked to contribute. An introductory section is concerned with the nature of aggression and war, and with popular beliefs about them. In successive sections the biological bases of aggression, the nature of individual human aggression, aggression within and between small groups, and international war, are discussed. In a final chapter the issues are brought together in a model that emphasizes their complexity.

We should like to acknowledge the work of Santiago Genovés, who inspired the project in the first instance; Martin Ramirez, who organized the Seville meeting; and David Adams, who has coordinated efforts for publicizing the Statement. We are also grateful to Ann Glover for the vital role she has played at all stages in the preparation of the manuscript.

J.G.
R.A.H.

THE SEVILLE STATEMENT ON VIOLENCE

Believing that it is our responsibility to address from our particular disciplines the most dangerous and destructive activities of our species, violence and war; recognizing that science is a human cultural product which cannot be definitive or all-encompassing; and gratefully acknowledging the support of the authorities of Seville and representatives of the Spanish UNESCO; we, the undersigned scholars from around the world and from relevant sciences, have met and arrived at the following Statement on Violence. In it, we challenge a number of alleged biological findings that have been used, even by some in our disciplines, to justify violence and war. Because the alleged findings have contributed to an atmosphere of pessimism in our time, we submit that the open, considered rejection of these mis-statements can contribute significantly to the International Year of Peace.

Misuse of scientific theories and data to justify violence and war is not new but has been made since the advent of modern science. For example, the theory of evolution has been used to justify not only war, but also genocide, colonialism, and suppression of the weak.

We state our position in the form of five propositions. We are aware that there are many other issues about violence and war that could be fruitfully addressed from the standpoint of our disciplines, but we restrict ourselves here to what we consider a most important first step.

IT IS SCIENTIFICALLY INCORRECT to say that we have inherited a tendency to make war from our animal ancestors. Although fighting occurs widely throughout animal species, only a few cases of destructive intra-species fighting between organized groups have ever been reported among naturally living species, and none of these involve the use of tools designed to be weapons. Normal predatory feeding upon other species cannot be equated with intra-species violence. Warfare is a peculiarly human phenomenon and does not occur in other animals.

The fact that warfare has changed so rapidly over time indicates that it is a product of culture. Its biological connection is primarily through language which makes possible the coordination of groups, the transmission of technology, and the use of tools. War is biologically possible, but it is not inevitable, as evidenced by its variation in occurrence and nature over time and space. There are cultures which have not engaged in war for centuries, and there are cultures which have engaged in war frequently at some times and not at others.

IT IS SCIENTIFICALLY INCORRECT to say that war or any other violent behavior is genetically programmed into our human nature. While genes are involved at all levels of nervous system function, they provide a developmental potential that can be actualized only in conjunction with the ecological and social environment. While individuals vary in their predispositions to be affected by their experience, it is the interaction between their genetic endowment and conditions of nurturance, that determines their personalities. Except for rare pathologies, the genes do not produce individuals necessarily predisposed to violence. Neither do they determine the opposite. While genes are co-involved in establishing our behavioral capacities, they do not by themselves specify the outcome.

IT IS SCIENTIFICALLY INCORRECT to say that in the course of human evolution there has been a selection for aggressive behavior more than for other kinds of behavior. In all well-studied species, status within the group is achieved by the ability to cooperate and to fulfil social functions relevant to the structure of that group. 'Dominance' involves social bondings and affiliations; it is not simply a matter of the possession and use of superior physical power, although it does involve aggressive behaviors. Where genetic selection for aggressive behavior has been artificially instituted in animals, it has rapidly succeeded in producing hyper-aggressive individuals; this indicates that aggression was not maximally selected under natural conditions. When such experimentally-created hyper-aggressive animals are present in a social group, they either disrupt its social structure or are driven out. Violence is neither in our evolutionary legacy nor in our genes.

IT IS SCIENTIFICALLY INCORRECT to say that humans have a 'violent brain'. While we do have the neural apparatus to act violently, it is not automatically activated by internal or external stimuli. Like higher primates and unlike other animals, our higher neural processes filter such stimuli before they can be acted upon. How we act is shaped by how we have been conditioned and socialized. There is nothing in our neurophysiology that compels us to react violently.

IT IS SCIENTIFICALLY INCORRECT to say that war is caused by 'instinct' or any single motivation. The emergence of modern warfare has been a journey from the primacy of emotional and motivational factors, sometimes called 'instincts', to the primacy of cognitive factors. Modern war involves institutional use of personal characteristics such as obedience, suggestibility, and idealism, social skills such as language, and rational considerations such as cost-calculation, planning, and information processing. The technology of modern war has exaggerated traits associated with violence both in the training of actual combatants and in the preparation of support for war in the general population. As a result of this exaggeration, such traits are often mistaken to be the causes rather than the consequences of the process.

We conclude that biology does not condemn humanity to war, and that humanity can be freed from the bondage of biological pessimism and empowered with confidence to undertake the transformative tasks needed in this International Year of Peace and in the years to come. Although these tasks are mainly institutional and collective, they also rest upon the consciousness of individual participants for whom pessimism and optimism are crucial factors. Just as 'wars begin in the minds of men', peace also begins in our minds. The same species who invented war is capable of inventing peace. The responsibility lies with each of us.

Seville, 16 May 1986

Signatories

David Adams, Psychology, Wesleyan University, Middletown (CT) USA; S. A. Barnett, Ethology, The Australian National University, Canberra, Australia; N. P. Bechtereva, Neuro-physiology, Institute for Experimental Medicine of Academy of Medical Sciences of USSR, Leningrad, USSR: Bonnie Frank Carter, Psychology, Albert Einstein Medical Center, Philadelphia (PA) USA; José M. Rodríguez Delgado, Neurophysiology, Centro de Estudios Neurobiológicos, Madrid, Spain; José Luis Díaz, Ethology, Instituto Mexicano de Psiquiatría, Mexico D. F., Mexico; Andrzej Eliasz, Individual Differences Psychology, Polish Academy of Science, Warsaw, Poland; Santiago Genovés, Biological Anthropology, Instituto de Estudios Antropológicos, Mexico

D. F., Mexico; Benson E. Ginsburg. Behavior Genetics, University of Connecticut, Storrs (CT) USA; Jo Groebel, Social Psychology, Erziehungswissenschaftliche Hochschule, Landau, Federal Republic of Germany; Samir-Kumar Ghosh, Sociology, Indian Institute of Human Sciences, Calcutta, India; Robert Hinde, Psychology, Cambridge University, UK; Richard E. Leakey, Physical Anthropology, National Museum of Kenya, Nairobi, Kenya; Taha M. Malasi, Psychiatry, Kuwait University, Kuwait; J. Martín Ramírez, Psychobiology, Universidad de Sevilla, Spain; Federico Mayor Zaragoza, Biochemistry, Universidad Autónoma, Madrid, Spain; Diana L. Mendoza, Ethology, Universidad de Sevilla, Spain; Ashis Nandy, Political Psychology, Center for the Study of Developing Societies, Delhi, India; John Paul Scott, Animal Behavior, Bowling Green State University, Bowling Green (OH) USA; Riitta Wahlström, Psychology, University of Jyväskylä, Finland.

Organizational endorsements

American Anthropological Association (Annual Meeting, 1986)
American Orthopsychiatric Association (1988)
American Psychological Association (Board of Scientific Affairs, Board of Social and Ethical Responsibility for Psychology, Board of Directors, and Council, 1987)
Americans for the Universality of UNESCO (1986)
Canadian Psychologists for Social Responsibility (1987)
Czechoslovak UNESCO Commission (1986)
Danish Psychological Association (1988)
International Council of Psychologists (Board of Directors, 1987)
Mexican Association for Biological Anthropology, (1986)
Polish Academy of Sciences (1987)
Psychologists for Social Responsibility (US 1986)
Society for Psychological Study of Social Issues (US 1987)
Spanish UNESCO Commission (1986/1987)
World Federalist Association (US National Board 1987)

(up to 31 August 1987)

A. AGGRESSION:
THE REALITY AND THE MYTH

1

The problem of aggression

ROBERT A. HINDE AND JO GROEBEL

In this chapter we address some general issues relevant to all the succeeding ones. These concern some matters of definition, the nature of aggressive behaviour, and its manifestation at different levels of social complexity.

Some terms used

This volume contains contributions from diverse disciplines. Not surprisingly, some of the terms used carry meanings differing according to the level of social complexity at which they are applied. More regrettably, at any one level there are sometimes differences in the ways in which the words are used. In this section we attempt to provide some precision not by attempting tight and ubiquitously applicable definitions, but by specifying the range of phenomena embraced by a number of commonly used terms.

Behaviour directed towards causing physical injury to another individual is labelled as *aggressive behaviour*: here there may be little disagreement. Aggressive behaviour refers more often to behaviour towards persons than to behaviour directed towards harming physical objects. It does not refer to behaviour that 'accidentally' results in harm to another. The harm referred to is primarily physical, though behaviour leading to psychological harm is often included.

However, a number of difficulties immediately arise. Although the notion of intent is central, the distinction between intentional and accidental harm to others is often difficult to make. Furthermore, the intent to harm may or may not be primary: in the latter case the harm caused may be instrumental to a further goal, such as the possession of an object or the acquisition of status.

In any case, aggressive behaviour often does result in settling status, precedent, or access to some object or space. Whether this is true of all aggressive behaviour is largely a matter of definition: in so far as an individual who causes physical injury to another is subsequently avoided, his aggressive behaviour has determined access to the space around him. The complementary question, of whether all behaviour that results in settling status, precedence or access is to be labelled as aggressive, gives rise to much disagreement. While many psychiatrists would include such behaviour, the authors in this volume would not. Behaviour that is merely assertive, and lacks the intention to inflict violence, is excluded. Phrases such as 'an aggressive salesman' confuse aggressiveness with assertiveness and, in our view, lead to unnecessary confusion.

Attack on another individual usually involves risk of injury for the attacker. It is therefore rarely single-minded, but is associated with self-protective and withdrawal responses. This is especially apparent in animal threat postures, many of which consist of a mosaic of elements of attack and threat, but is also to be seen in the boxer's stance and much human threat and bombast. Because of the close association between them, some workers lump together attack, threat, submissive and withdrawal behaviour as *agonistic behaviour*, even though some types of behaviour in this category are clearly aggressive and others are not.

What has been said so far refers primarily to behaviour, but dreams, fantasies and feelings involving the intentional inflicting of harm on others are also properly described as aggressive. It will be apparent that the category of aggression or aggressive behaviour, whilst containing a generally agreed nugget, is shady at the edges.

Some additional terms are related to the concept of aggression:

Violence implies the infliction of physical (but usually not psychological) harm on another individual or an object. It usually involves harm that is fairly severe. The harm is usually, but not necessarily, intended.

Conflict in a broad sense refers to a conflict of interest or a disagreement over status or the allocation of resources. Conflict may occur between individuals or groups, and usually implies that the situation is perceived by the participants as one in which, if one wins, the other will lose. *Conflict resolution* can occur in a large number of ways, and does not necessarily involve aggression. *Competition* is limited to situations in which the resource in question is limited. Competition can occur without the participants being aware of the conflict, as when two individuals search independently for a limited resource.

War is a special type of aggression. It involves aggression between groups in which the individuals are in some degree organized towards achieving the common goal. It is usually institutionalized, with individuals occupying distinct roles (soldier, general, munitions worker, etc. – see below).

Categories of aggression

While for many purposes it is convenient to group together as aggressive behaviour all instances in which individuals or groups direct their behaviour so as to harm others, such a category conceals considerable internal heterogeneity. This is reflected in the numerous attempts that have been made to classify aggressive acts into distinct sub-categories. For example, in studies of childhood aggression, four categories are often recognized. These are related to the underlying motivations (Feshbach, 1964; Manning, Heron & Marshall, 1978):

1 *instrumental* or *specific aggression*, concerned with obtaining or retaining particular objects or positions or access to desirable activities;
2 *hostile* or *teasing aggression*, directed primarily towards annoying or injuring another individual;
3 *defensive* or *reactive aggression*, provoked by the actions of others;
4 *games aggression*, involving deliberate attempts to inflict injury escalating out of physical games.

Again, one system for classifying aggressive violence by adolescents involves the following categories (Tinklenberg & Ochberg, 1981):

1 *instrumental*;
2 *emotional*;
3 *felonious*, committed in the course of another crime;
4 *bizarre*, involving psychopathic behaviour;
5 *dyssocial*, acts that gain approbation from other group members or from the reference group.

Other systems have included phenomenological dimensions such as verbal vs physical aggression, or target dimensions such as aggression directed towards persons vs animals vs objects.

Such attempts at categorization often provide clarification. For instance, the different types of childhood aggression have different developmental courses and probably differ in their prognostic value for adolescent behaviour problems. However, no system is wholly satisfactory: instrumental aggression by children is not always clearly distinguishable from hostile aggression, which may be related to long-term goals concerned with access to desirable objects; and felonious violence may overlap with both instrumental and emotional violence.

These problems reflect the fact that acts of aggression can be seen as involving more than one type of motivation. In addition to the propensity to harm others they may involve fear (as discussed above), acquisitiveness and/or assertiveness (especially in the case of instrumental aggression) and no doubt sometimes other types of motivation (e.g. sexual). The relative strengths of these different motivations may vary from case to case, making overlap inevitable in any attempt to categorize acts of aggression.

Levels of social complexity

The aetiology of aggressive behaviour differs markedly with the social complexity of the situation in which it occurs, and it is convenient to have a conceptual framework against which to consider these differences. Aggressive behaviour (including war) involves behaviour by *individuals*, and thus necessarily depends on behavioural mechanisms within individuals. Many episodes of aggression involve an *interaction* between two individuals, an interaction referring to a series of exchanges occurring within a limited span of time. The nature of each interaction will be affected by the characteristics of both the individuals involved.

When two individuals have a series of interactions, each interaction may be influenced by the preceding ones and/or by expectations of future ones. We then speak of them as having a *relationship*. We tend to classify relationships according to the content of the interactions involved: thus, a mother–child relationship differs from a teacher–pupil relationship because the individuals concerned do different things together. But within any one type of relationship the interactions may vary in content, quality, intimacy and in a number of other ways. The nature of any particular relationship depends on the nature and patterning of the interactions of which it is composed. At the same time the nature of a relationship affects the nature of its constituent interactions, because the individuals concerned guide their behaviour according to their experience within and expectations for that relationship. And in the longer term the behaviour individuals *can* show is affected by the relationships they have experienced.

Relationships are set within networks of other relationships, such as family groups, work mates and so on. Each relationship is affected by the nature of the *group*, and the nature of the group depends upon the constituent relationships.

It is thus convenient to consider a succession of levels of social complexity – individuals, interactions, relationships and groups – with two-way causal relations between them. Of course, the distinctions are sometimes blurred: interactions grade into relationships and a triadic relationship is already a group. More important, the levels are to be considered not as entities but as processes in continuous creation by the dialectical relations between them (Hinde, 1979, 1987).

Since the behaviour of individuals is affected by their beliefs, values and so on, and these are to some extent shared by the individuals in a group, it is convenient to consider also another level, that of the *socio-cultural structure*, superimposed on all of these. This includes not only beliefs, values and myths, but also the institutions with their constituent roles recognized in the society. Most societies have many institutions, with rights and duties associated with each role within them. For instance, marriage (with the roles of husband and wife), and Parliament (with the roles of President or Premier, the members, the voting public and so on) are institutions of most western societies. For each institution there are generally accepted norms and values as to the proper way for the incumbents of each role to behave. The behaviour of each individual is thus affected by common behavioural propensities (e.g. in the case of marriage, to behave sexually and parentally), by idiosyncratic aspects of their own personality, and by the rights and duties associated with the role currently occupied (e.g. in this case perhaps a new residence, special ways of dressing, a reorientation of sexuality, and so on). The socio-cultural structure embraces not only the institutions, beliefs, etc., but also the relations between them. It can be seen as objective, something that could be abstracted from a participant observer's description of the society, or as subjective, existing in the minds of the various members of the society, perhaps in a slightly different form in each individual. It is related by dialectical relations with each of the levels of social complexity we have discussed.

The value of this conceptual framework can be exemplified by a brief discussion of three types of aggressive behaviour. First, consider two boys fighting over a toy. Each boy is acquisitive and attempts to gain the object, using aggressive behaviour to that end. Each may also hope to enhance his status in the peer group by overcoming his rival. The motivations involved, though multiple, are relatively simple.

Now consider a fight between two groups only recently touched by civilization. Each participant stands to gain, in booty, wives, or what have you, as well as in prestige among peers. Thus, aggression is used as a means to an end by acquisitiveness and assertiveness. But institutions also play a role. Individuals may be encouraged to fight and even to sacrifice themselves by religious leaders who inculcate beliefs about rewards in another life for those who die in battle, the approval of ancestors, and so on. Thus, the motivations of the individual combatants stem both from basic propensities and from the dictates of religious (and perhaps other) institutions in their society.

In modern warfare the importance of institutions is even greater. Indeed, modern warfare is itself properly considered as a human institution with a wide variety of constituent roles. Among the combatants, aggressive propensities play little part. Cooperation with others and tendencies to obey superiors are certainly crucial. Primarily, however, they fight because of their beliefs. They believe that by so doing they can protect individuals and values that they hold dear. They may even go more readily into battle because they believe it will lead them to a better after-life. These beliefs and values are inherent in their culture and are accentuated by governmental propaganda as part of the institution of war.

But the institution of modern war has many roles in addition to that of the fighting man, e.g. generals, political leaders, munitions workers, medical workers, and so on. Each of these has special roles and duties assigned to them. In many cases the duties contribute to the maintenance of the institution by stimulating or reinforcing the belief systems of others. In every individual, whatever his role, lack of patriotism may already be cause for mistrust.

There are thus tremendous differences between individual aggression and war in the factors influencing the behaviour. At every level, the beliefs held by individuals are important, but in the case of institutionalized war they become crucial. If we are to understand the bases of either individual aggression or war, therefore, we must come to terms with what individuals believe about it. This is the subject of the next chapter.

References and further reading

Brain, P. F. (1984). Biological explanations of human aggression and the resulting therapies offered by such approaches. *Advances in the Study of Aggression*, 1984, 63–102.

Feshbach, S. (1964). The function of aggression and the regulation of aggressive drive. *Psychological Review*, **71**, 257–62.

Hamburg, D. A. & Trudeau, M. B. (eds) (1981). *Biobehavioral Aspects of Aggression*. New York: Liss.

Hinde, R. A. (1979). *Towards Understanding Relationships*. London: Academic Press.

Hinde, R. A. (1987). *Individuals, Relationships and Culture: ethology and the social sciences*. Cambridge: Cambridge University Press.

Johnson, R. N. (1972). *Aggression in Man and Animals*. Philadelphia: Saunders.

Karli, P. (1987). *L'Homme Aggressif*. Paris: Odile Jacob.

Manning, M., Heron, J. & Marshall, T. (1978). Styles of hostility and social interactions at nursery, at school and at home. An extended study of children. In L. A. Hersov & D. Shaffer (eds), *Aggression and Anti-social Behaviour in Childhood and Adolescence*. Oxford: Pergamon Press.

Tinklenberg, J. R. & Ochberg, F. M. (1981). Patterns of violence: a California sample. In D. A. Hamburg & M. B. Trudeau, (eds), *Biobehavioral Aspects of Aggression*. New York: Liss.

2

Beliefs about human aggression[1]

JEFFREY H. GOLDSTEIN

What we believe about human behavior determines how we act, both as individuals and as nations. In the world of human affairs, objective truth often matters less than subjective reality. If we believe something to be true, we behave as though it were, and sometimes act to ensure that it is. Scientists from many disciplines have studied aggression in humans and other species in an attempt to understand the bases and essential nature of human aggression. Many summaries of scientific research on aggression are available. In this chapter, however, the concern is not with what social and behavioral scientists *know* about human aggression, but what many of us, layman and scientist alike, *believe* about it. The two are rarely the same.

Beliefs about aggression

Historical epochs differ in their views of human nature. Whether aggression is believed to be an inevitable part of human behavior has changed from time to time. Since the mid-twentieth century, the following beliefs about human aggression seem to have prevailed in Western Society:

1. Humans are instinctively aggressive. Safe, acceptable channels for the expression of this destructive instinct must be found or society as a whole will suffer.
2. Failure to express anger results in heart disease, stress, and high blood pressure. Expressing anger is a healthy way to deal with it.

---

[1] This chapter is adapted from Chapter 1 of *Aggression and Crimes of Violence* (second edition), Oxford University Press (1986), and a chapter in J. Martin Ramirez, R. A. Hinde, & J. Groebel (eds), *Essays on Violence*, University of Seville Press (1987).

3 The aggressive instinct can be controlled through substitute activities, such as football games and Sylvester Stallone movies and expressing hatred toward an enemy. These enable us to rid ourselves vicariously of the urge to commit destruction.

4 Children should be allowed to play aggressively. This will 'get it out of their system' and they will be better behaved as a result.

5 Extreme acts of violence, such as terrorist acts, child and spouse abuse, are usually committed by individuals who are mentally ill.

6 Violence is the result of the aggressive drive. In some individuals and groups this motive is abnormally strong.

7 War is an expression of the aggression instinct. It is unavoidable because humans have an inborn need to satisfy their aggressive urges. Peace is an aberration, a temporary period between wars. War has always been, and hence will always be, with us.

These statements constitute part of what I have called the *mythology of aggression* (Goldstein, 1986*a*). They are widely believed, but evidence on their behalf is lacking or inconclusive. These myths form the basis for many of our beliefs about human violence, crime, and war. They also color our efforts to deal with violence at home, with crime in the streets, and with policy toward other countries. To the extent that our dealings with others are based on these beliefs, our actions are apt to prove ineffective and in some instances to produce the very conditions they are designed to ameliorate.

Is aggression an instinct?

Throughout modern history the answer to the question of whether aggression in humans is innate or learned has swung first one way, then the other. Because at various times the answer to the question has changed, there exist in our social, legal, and political systems and in our foreign policies methods and remedies for the control of violence that are not necessarily consistent with one another nor with current scientific thinking on the issue.

The instinctual bias that underlies many beliefs about human aggression can be attributed to five influences:

1 the accounts of aggression by some ethologists and sociobiologists, such as Konrad Lorenz (1966) and others;

2 the dramatic research on electrical and chemical stimulation of the brain;

3 the popularity and pervasiveness of Freudian theory (1955);

4 the idea that aggression is in our genes;

5 the biological emphasis in the mass media's presentation of violence and aggression research (Goldstein, 1986*b*).

One giant leap towards mankind?

The ethological arguments proposed by Lorenz and others reasoned that there is ample evidence that our animal ancestors were instinctively violent beings, and since we have evolved from them, we too must be the bearers of destructive impulses in our genetic make-up. Lorenz states 'There cannot be any doubt, in the opinion of any biologically minded scientist, that intrapyschic aggression is, in man, just as much of a spontaneous instinctive drive as in most other higher vertebrates.' (1964, p. 49).

In his book, *On Aggression*, Lorenz (1966) argues that while humans and animals share an instinct for aggressive behavior, humans, unlike other species, lack a well-developed mechanism for the *inhibition* of modern means for aggression. Many animal species inhibit aggression in response to the visible and audible pain and suffering of their victims. Human technology has made possible the infliction of injury and death at distances too great to enable perception of these 'pain cues', so we humans do not inhibit our aggression at the sound and sight of our enemies' suffering.

There are essentially two difficulties with these arguments. First, the evidence that animals, at least the higher primates, are instinctively aggressive is not very convincing. This issue is discussed in later chapters. Second, even if the evidence reviewed by some ethologists and sociobiologists were sufficient to establish that infrahuman species are innately violent, we would still have to ask whether that proves anything at all about proneness to aggression in humans. Of course, it does not.

There is not much doubt that humans *can* behave like phylogenetically inferior species. There is no reason why an organism with a complex nervous system, such as ours, cannot behave like or mimic the behavior of animals with less complex nervous systems. However, to argue that, because we *can* behave like lower organisms we *must* behave like them, is fatuous. If we take learning experiments on animals, such as operant conditioning studies in which an organism is rewarded for performing a certain response, there is no neurological or psychological reason why a human being cannot also learn responses in this way. To argue that this is the way humans learn – simply because it is one way in which they may learn – begs the question. It is obviously true that humans can and do learn behaviors through operant conditioning. But there may also be species-specific means of learning (or of aggressing) in humans.

In the brains of primates

The most dramatic demonstrations of the role of the limbic system in aggression were presented by José Delgado (1969). By implanting radio receivers in the brains of cats, monkeys, and other species, he has been able, within limits, to control the aggressive behavior of his research subjects by stimulating the hypothalamus.

Perhaps the most frightening implication of this research is the possibility that human aggression could be manipulated without the actor's awareness or consent. Electrodes implanted in the brain of a newborn infant or of a surgery patient could be controlled by the possessor of the appropriate radio transmitter. A number of authors have made eloquent appeals to limit such potentially antisocial technologies.

In many of the brain stimulation studies on aggression, it has been reported that even when stimulated some animals will not engage in aggression unless there is available an 'appropriate' target to attack. For example, if the only available target is a more dominant animal than the one stimulated, no attack will occur. Only when the target is inferior in status will electrical stimulation reliably result in attack. Thus, even with infrahumans, brain stimulation does not guarantee that aggression will ensue.

But let us suppose the research did find that stimulation of certain areas of the hypothalamus could reliably elicit aggression from humans as well as from other animals. What would this tell us about human aggression? Even if stimulation of the limbic system is a necessary condition for human aggression, we would still need to determine how social and psychological events trigger the physiological reactions that result in activation of the hypothalamus. In any case stimulation of particular brain areas does not guarantee that aggressive behavior will follow. It is likely that in natural situations – those of non-intervention in the brain – it is cognitive and environmental factors that are themselves responsible for the stimulation in the first instance. It is likely that those portions of the brain found during surgical intervention to elicit aggression are not primarily involved in natural situations.

Genes and the genesis of aggression

Those who make the case that aggression is biologically caused often attach their arguments to a particular finding that becomes faddish.

Western society has often subscribed to the belief that those who *act* differently must *be* different. For example, at the turn of the century, criminals were believed to be of a certain physiological 'type'. The criminologist, Cesare Lombroso, wrote:

> The criminal by nature has a feeble cranial capacity, a heavy and developed jaw, projecting [eye] ridges, an abnormal and asymmetrical cranium . . . projecting ears, frequently a crooked or flat nose. Criminals are subject to color blindness; left-handedness is common; their muscular force is feeble. . . .
>
> (Quoted in S. Chorover (1979), pp. 179–80.)

In arguing that criminality could be predicted from physiognomy, a 'rational criminology' became possible. In the late 1960s and 1970s the belief that hyperaggressiveness in males was caused by a genetic abnormality was widespread. It had been reported that males imprisoned for violent offences were more likely to have an extra Y chromosome, the so-called 'XYY males'. The incidence of XYY males in the general population is approximately 1 in 3000. However, it was found that approximately 3 per cent of the men in maximum security prisons and hospitals for the 'criminally insane' in Edinburgh were XYYs.

Subsequent studies and critiques cast doubt on the importance of the additional Y chromosome in criminality. Of course, over 97 per cent of those imprisoned for violent crime were not XYY men, nor was it clear what proportion of XYY males did not show signs of antisocial conduct. Furthermore, intelligence is likely to mediate the relationship between genes and detected crime in that the least intelligent men were apt to be caught and convicted. Montagu (1968) has pointed out that the XYY phenomenon is not a syndrome that reliably leads to particular behaviors, but an anomaly that, depending upon environmental conditions, may lead to a host of possible behaviours (see chapter by Manning in this volume).

For decades, paleoanthropologists and biologists believed that stone artifacts found with ancient humanoid skeletal remains were destructive weapons. The existence of such 'weapons' was taken as evidence of our early ancestors' innate aggressiveness. However, within the last few years, some anthropologists have found it more plausible to believe that these objects were not weapons at all but tools used to scavenge for food. Time and again efforts to find a tangible, material basis for human failings have themselves failed.

Freudian slips

A major influence on twentieth century thinking about human nature is psychoanalytic theory. This is not the place, and I am not the person, to undertake an overview and critique of the theory. Freud's fascination with aggression and the development of his ideas about human violence have been well documented.

Psychologists have often misinterpreted or misunderstood Freudian aggression theory. Whatever their understanding of Freud may be, many of the behavioral predictions they derived from psychoanalytic theories of aggression are untestable or are not supported by quantitative data.

The general biological model underlying psychoanalytic theories of aggression is crude and partly false. Freud considered aggression to be an instinct which, like hunger, is constantly building up unless satisfied in reality or in fantasy. Not only is this not the way that aggressive behavior operates, it is not even how hunger works. But the theory provides us with something that has often proven to be more useful than biological truth: a justification for human violence. If humans are instinctively aggressive beings, then there is no use denying what nature has seen fit to provide.

The biological model also implies that human, like much animal, aggression is highly stereotypic, consisting of characteristic forms of expression. However, anthropologists have long written of the flexibility and cultural relativity of aggression, and as early as 1939 three prominent psychologists, Boring, Langfeld & Weld, could write (p. 163):

> Conflict between individuals does not invariably or universally result in the same behavior. Instead of fighting with his fists, the Kwakiutl Indian fights with property in the institution of the 'potlatch', in which the more property he can give away or destroy, the more superior he is to his opponent. Eskimos settle their conflicts in a public contest in which each sings abusive songs about the other. When two Indians of Santa Marta quarrel, instead of striking each other they strike a tree or rock with sticks, and the one first breaking his stick is considered the braver and hence the victor. In other societies aggression is expressed in still other ways; even within the same society there may be a wide range of different socially approved expressions of aggression.

In many societies aggression is rare or absent.

This means that, whatever the bases of human aggression, it is within the capacity of humans to do away with it.

What the media tell us about violence

At the Fifth Biennial Meeting of the International Society for Research on Aggression, held in Mexico City, nearly 100 scholarly papers were presented. These covered the broad range of aggression, violence, and peace research: human and animal brain studies, a model for predicting war presented by a distinguished political scientist, studies on personality and violent crime, and anthropological analyses of the control of aggression. From these many varied and interesting papers, only one was written up in the English-language press, and that was a study of an experimental drug that inhibits mouse-killing in rats. The political, social, and psychological bases of violence were not mentioned in the press.

In order to examine the underlying model of human violence that might result in such selective reporting of research, I began collecting all articles on violence, aggression, and crime research appearing between 1982 and 1984 in three leading American newspapers (*Los Angeles Times, New York Times, Washington Post*) and three news magazines (*Newsweek, Time, US News & World Report*). The focus of our analysis was on the underlying model of human violence that might result in the selective reporting of research. While content analysis of these articles is not yet complete, I have several distinct impressions from reading scores of such articles. Foremost among them is the *mechanistic view* of behavior implicit in journalistic accounts of research on violence and crime. Although some of the articles discuss research on social and psychological factors involved in violence, there is a consistent theme:

1 It is assumed that the causes of human violence exist within the individual.
2 It is also assumed that if psychiatrists, psychologists, and biologists were only clever enough, they could identify the genetic or personality factors that give rise to violent behavior.
3 Given the 'fact' that the causes of violence reside within the individual's skin, it is assumed to be at least theoretically possible to identify potential offenders before they ever commit an offence by using some sort of bio-genetic or psychological screening procedure.
4 A good many psychologists, psychiatrists, anthropologists, and other scientists also employ these assumptions.

The hard evidence for this 'model' of human aggression is lacking.

What are the implications of such a perspective? One is to reinforce the belief that those who commit acts of violence are different in tangible and predictable ways from those who do not. Second, this view also undermines the actor's own sense of efficacy and social responsibility by suggesting that he is impelled to act antisocially. There is a self-fulfilling prophecy in this view of criminality. Many individuals who commit violent crimes hold the same beliefs about aggression as the rest of the population (having learned them from the same sources, notably the mass media). They also believe, and frequently argue in court, that their crime was uncontrollably caused by a physical or psychological (rather than a personal or moral) defect.

Why do we so often pin our hopes for explaining violence on a biological, genetic, or physiological entity? Perhaps the primary reason for this reductionist emphasis is an attempt to separate ourselves (who act in an acceptable manner) in a tangible way from those whose behavior we deplore. We can feel comfortable, satisfied, and relatively safe knowing that we do not possess the 'disease' of antisocial behavior. We also provide ourselves with a ready-made excuse should we exceed the bounds of propriety.

If we convince ourselves that the lawless, the violent, and the opposition are qualitatively different, we are then in a position to treat them expediently, since they are not completely human or completely 'normal'. In the case of nations, we tend to attribute what we see as their aggression to some inherent characteristic of their government or belief system. And because of this defect, they are never to be trusted. If we ourselves commit an act of violence, then our own individual responsibility for it can be minimized to the extent that our behavior was 'impelled' by tangible forces beyond our control.

It is tempting to adopt the reductionist position, not only because it allows us to be complacent, but because we have so often been exposed to it. In countless tales, films, books, and television programs we have seen individuals whose violence was the result of brain damage, schizophrenia, physical agents such as drugs, alcohol, and 'bad genes'. We have seen our enemies – whoever they may be at the moment – portrayed as inherently evil, primitive and irrational. Social and psychological causes of violence, which are abstract and complex and therefore more difficult to portray, are less often found in public accounts of crime, aggression, and war.

Scientists, no less than others, tend to be reductionist in their belief that aggression's ultimate explanation will be biological. Because sound methodology is difficult to achieve in the study of intangible causes of behavior, theory development suffers. Since theory on the origins of human violence is relatively weak, scientists tend to take the path of least resistance and ascribe aggression to biological necessity. The problem is thus 'solved'. For example, Freud initially attempted to explain human aggression in terms of the life instinct, but he was unable to account for the atrocities of World War I in this way. He then proposed an aggression instinct and events became more readily 'understandable'. Another reason scientists tend to reject social and psychological explanations for violence is that many of these theories have been so simple and one-dimensional. If one tries to account for aggression in purely sociological terms, such as social class and status deprivation, there are counterexamples that cannot easily be fitted within the theory. For example, some have argued that interpersonal violence is found most often among the poor and powerless, but they immediately encounter the fact that most people who are poor and feel powerless are not aggressive, despite the fact that most aggression may be committed by individuals fitting such a description. Clearly, additional factors must be invoked to account for why only some poor and powerless people are aggressive while the majority is not. These factors have so far eluded sociologists. Of course, counterexamples do not necessarily indicate that a theory is false, only that it may be too simple. It is, in fact, desirable in science to arrive at the simplest theory permitted by the data, and scientists often resort to the simplest of all possible theories, that people behave the way they do because they are incapable of doing otherwise.

Human cognition and aggression

Humans are the earth's supremely symbolizing creatures. We are able to impart meaning of near infinite variety to any object or action. A piece of multicolored cloth may evoke feelings ranging from disinterest to nationalistic fervor, depending upon whether it is a sample of fabric or the flag of the Third Reich. We can find in a small gesture or subtle glance worlds of meaning, capable of eliciting feelings, thoughts, and actions far in excess of such barely noticeable stimuli.

Certainly one of the distinguishing characteristics of our species is this symbolic capacity. It enables us to represent the most complex of ideas using only marks of ink on a page. It makes possible the planning of a trajectory for a retrievable space vehicle, but also the planning of a trajectory for a nuclear weapon. If humans as a species appear always to have been more destructive than their circumstances warrant, it is not because they have been bequeathed excess genetic baggage from their primate ancestors. Rather, it is because the very capacities that allow us to shape our environment to suit our convenience also allow us to perceive threats and enemies where none exists. We can even imagine on the basis of barely tangible artifacts that other species are innately violent. There has never been destruction of the human type or scale in any known species. We did not obtain our aggressiveness through those genes we share with other species. I believe we obtained our capacity for violence of the human sort along with our ability to reason abstractly, to see things, like gravity or threat, that are not materially there. We are the only species capable of aggressing because of the beliefs we hold.

If there is any truth to this argument, then human violence of any sort is not inevitable, and certainly it is not beyond our ability to control. If cognitions can 'override' the body's need for nutrition, as in a hunger strike, then cognition can override whatever propensities for a nuclear strike may exist among us.

References and further reading

Boring, E. G., Langfeld, H. S. & Weld, H. P. (1939). *Introduction to Psychology*. New York: Wiley.

Chorover, S. (1979). *From Genesis to Genocide*. Cambridge, Mass.: MIT Press.

Delgado, J. M. R. (1969). *Physical Control of the Mind*. New York: Harper & Row.

Freud, S. (1955). Beyond the pleasure principle. Vol. 18. In J. Strachey (ed.), *The Standard Edition of the Complete Pyschological Works of Sigmund Freud*. London: Hogarth.

Goldstein, J. H. (1986a). *Aggression and Crimes of Violence* (second edition). New York: Oxford University Press.

Goldstein, J. H. (1986b). *Reporting Science: the Case of Aggression*. Hillsdale, New Jersey: Lawrence Erlbaum.

Lorenz, K. (1964). Ritualized fighting. In J. Carthy & E. Ebling (eds), *The Natural History of Aggression*. New York: Academic Press.

Lorenz, K. (1966). *On Aggression*. New York: Harcourt, Brace & World.

Montagu, A. (ed.) (1968). *Man and Aggression*. New York: Oxford University Press.

Ramirez, J. M., Hinde, R. A. & Groebel, J. (eds) (1987). *Essays on Violence*. Seville: University of Seville Press.

B. BIOLOGICAL MECHANISMS IN THE INDIVIDUAL

Editorial

Does what we know about the nature of aggressive behaviour and its underlying mechanisms provide any evidence to support the view that there is a certain inevitability in human aggression? This is the theme underlying the chapters in this section.

In general, the questions that can be asked about behaviour and its underlying mechanisms fall into four groups, concerning respectively immediate causation (What elicited it? How does the mechanism work?), its development (What was the nature of the interaction between organism and environment, from the zygote on, that led to the present condition?), its evolution (Can we trace the evolution of the behaviour or its underlying mechanisms in our animal ancestors?), and its function (Through what beneficial consequences did natural selection act to favour its maintenance?). These questions form the background to the chapters in this section.

Huntingford is concerned with the relations between animal and human aggression. Many but by no means all animal species show some form of aggressive behaviour, but it is always flexible, influenced by past experience and adjusted to the current situation. Few show organized group aggression and there is nothing approaching modern war. Individual aggression has often been ascribed to an aggressive 'instinct', but Bateson shows that instinct is a dangerous concept embracing a number of unrelated issues. Nor, as Manning shows, is human aggression inevitable by virtue of our genetic constitution. Certainly, aggressiveness is influenced by genetic constitution, but that says nothing about inevitability. And the fact that individuals can behave aggressively indicates the presence of the neural and endocrine machinery for aggression but, as Herbert argues, whether or not it is called into use depends on a host of experiential and contemporaneous environmental factors.

Perhaps the overall message from the chapters should be the danger of simple either/or questions. Yes, one can trace evolutionary influences on our aggressive behaviour, but it is not determined by them. No, it does not make sense to ask whether aggressive behaviour is driven from within or elicited, innate or learned. Yes, aggressive behaviour is influenced by our genes, but it is not determined by them. Yes, we do have mechanisms for aggression, but they need not be called into play. While the job of behavioural and social scientists is to find simplifying regularities in the complexity of human behaviour, the posing of too-simple questions prevents progress. And that is nowhere more true than in the understanding of aggression and war.

3

Animals fight, but do not make war

FELICITY ANN HUNTINGFORD

Current public discussion about human aggression and warfare is often influenced by views of the biological bases of aggression that have long been discarded by behavioural scientists. These views have implications about whether violence can be controlled, and how, and it is therefore important to set the record straight.

One such view is that the human habit of hunting and killing large animals inevitably brings with it high levels of aggression towards other human beings. This view is wrong: fighting between members of the same species and predatory attack by a carnivorous animal on its prey have different causes and different functions, such that there is no necessary association between hunting and aggression. Thus, many carnivores live peaceful lives, while plant eaters such as the mountain sheep are among the most aggressive of all animals when it comes to fighting rivals of the same species. The roots of human aggression do not lie in the hunting habits of our ancestors.

A further influential but now discredited view on the nature of human aggression is typified by the following quotation from Konrad Lorenz's 1966 book *On Aggression*:

> To the humble seeker after biological truth there cannot be the slightest doubt that human militant enthusiasm evolved out of a communal defense response of our pre-human ancestors. The unthinking single-mindedness of the response must have been of high survival value ... our only hope of ... (controlling this powerful motivating instinct) rests on humble recognition of the fact that militant enthusiasm is an instinctive response with a phylogenetically determined releasing mechanism. ...

At the time Lorenz was writing, the available evidence seemed to some to support this view. Now, after 20 years of further research on the subject, our picture of how behaviour is organized and about the way natural selection acts on it is very different.

According to now-discredited instinct theory, humans have inherited from their animal ancestors an internally generated tendency to fight. During interactions between human groups, the aggressive drive manifests itself as warfare. The drive builds up inside us until it reaches a critical level, when unthinking, irrational attacks are elicited by any individual possessing certain critical features (key stimuli). The act of attacking dissipates the aggressive drive, but peace lasts only until it builds up again, as it necessarily must. It is argued that this condition has evolved because those individuals among our ancestors who fought readily and fiercely, regardless of risk, were favoured by natural selection over those who hesitated. Because, according to the instinct theory, aggression is inherited, internally generated and triggered by simple stimuli, it cannot be prevented by controlling our environment. Our best hope is to direct it into useful, or at least harmless, channels.

As discussed by Bateson elsewhere in this book, theories of this type have a long history. Because they are still so influential, the present chapter considers the ways in which they fail to meet the tests of modern scientific research. The validity of the instinct approach and its use as an explanation of human warfare depends on three issues:

1 Do the mechanisms that control aggression and its effects on fitness really operate as instinct theory claims?
2 If so, does this mean that human aggression is organized in the same way? (In other words is it legitimate to apply conclusions derived from what animals do to the behaviour of our own species?)
3 If this is the case in general, do any non-human animals show behaviour that is equivalent to human warfare?

The causes and functions of animal aggression

Aggression of some sort is widespread (though not universal) among animals, and many species possess specialized weapons that are deployed during fights. For example, sea anemones drive away other individuals that approach too close by beating at them with club-like collections of poison cells, while rag worms use their proboscis to expel intruders from their burrows, funnel web spiders fight over well-placed feeding sites, male cichlid fish wrestle with locked jaws over places to breed, and male gladiator frogs use sharp spines on their thumbs during grappling combat over mates. Similarly, male ruffs struggle for places from which to display to females and red deer stags fight to sequester receptive females during the rut. Many animal fights take place (directly or indirectly) over valuable resources that are in short supply; the winner of a fight may get exclusive or preferential access to food, shelter or mates. Thus, there is nothing abnormal or unnatural about aggression; rather it is part of the repertoire of behavioural responses that animals use to cope with their environment.

However, neither the mechanisms that control aggression nor its functions resemble those postulated by instinct theory. Internal factors (such as certain hormones and brain chemicals) can facilitate aggression, but these do not inevitably accumulate in peaceful animals. Even when animals are motivated to fight, attack is not a reflex response to a simple sign stimulus; whether or not two potential opponents come to blows depends on a complex assessment of their probability of winning or of being injured and of the value of the disputed resource. A great deal of scientific effort has recently been aimed at analysing the way risk of injury devalues the benefits of winning fights. We now know that, while animals who fight fiercely regardless of circumstances may win more fights, those that adjust their aggressive behaviour to its likely costs and benefits are more successful in the long run.

Although the winner of a fight often gains access to valuable resources and although some sort of aggression is found in almost all groups of animals, really destructive fighting is relatively rare. Fights usually start with an exchange of displays (spiders raise and vibrate their walking legs, fish raise their gill covers, red deer stags roar and strut about) which may then escalate to actions involving some degree of physical contact (such as tail-beating and mouth-wrestling in fish). Damaging fights in which weapons are used in earnest occur only after this stage, although animals can and do injure each other in fights when the rewards are sufficiently high. For example, mantis shrimps fight to the death over shelters when these are necessary to avoid desiccation, while fights between red deer stags during the rut often escalate, with many stags receiving serious injuries. However, fights are usually resolved (by one animal accepting defeat and retreating) before escalation is complete and thus before any damage has been done to either participant.

Escalated, injurious fighting is relatively uncommon because, when two potential opponents meet, instead of reacting blindly to the key feature presented by their rivals (as depicted by the instinct theory) their behaviour is determined by a complex of factors relating to the chance of victory and injury and the rewards for winning. At one time it seemed that attack in many animals was triggered by just one or a few specific stimuli from an opponent (the red belly of a male stickleback being the best known example), with other features (such as its size and shape) being irrelevant. More recent evidence shows that this is not the case, and that fighting animals continually adjust their behaviour to many aspects of the appearance and behaviour of their opponent. A larger rival, a relative, a previously victorious opponent or one that shows signs of aggressiveness is less likely to be attacked than one showing the opposite features. In addition, behaviour during fights is modified by the context in which the encounter occurs. Many animal species defend their territory against intruders of the same species. In such territorial species, an animal fights less fiercely away from its home ground and when defending a poor-quality patch.

The complicating factors that make aggression more than a blind response to a releasing stimulus are particularly important among long-lived, intelligent social animals, such as the primates, our nearest living relatives. Most primates live in relatively permanent groups made up of a number of animals of different sexes and ages. This means that the same individuals may come into conflict repeatedly and that dominance–subordinance relationships are established. In such a case, whenever two individuals meet, one (the subordinate) regularly gives way to the other (the dominant). In a social group, the result of a number of such two-way relationships may be a more-or-less linear hierarchical organization, with each group member having its own status, ranging from the alpha or dominant individual at the top to the omega, subordinate individual at the bottom. Some sort of dominance organi-zation is found in most primate groups, but the complex social inter-actions that occur in primates mean that status is not determined by fighting ability alone and that resources are not distributed simply by status.

The relationship between a dominant and a subordinate primate is not just based on the exchange of aggressive actions. Patterns of attention and social affiliation, expressed commonly in the form of grooming, are also important. In addition, two or more group members often form alliances, so that subordinates can collectively overcome a dominant animal. The members of a primate social group are often related to each other and alliances are usually formed between related individuals. However, a baboon, for example, may go to the assistance of an unrelated animal by whom it has itself been aided in the past. Status in primate groups therefore has as much to do with intelligence, social skills, kinship patterns and the ability to cooperate as it has to do with aggression, and the dominant animal in a troop is not usually the most aggressive. In some circumstances, dominant primates derive benefit from their high status; for example, high-ranking male baboons perform most matings with females at the peak of their fertility and high-ranking female vervet monkeys get first access to water holes during droughts. However, on a day-to-day basis, subordinate animals are not deprived of the necessities of life.

Extrapolating to humans

The justification for applying what we know about animals to our own species lies in the fact of evolution. We are part of the animal world and have inherited many traits from our non-human ancestors; these we can expect to share particularly with our nearest living relatives, the Old World monkeys and apes. There is ample evidence that this is the case for many morphological traits (a flexible hand with opposable thumb, for example) and for physiological ones (such as the role of insulin in the control of glucose metabolism). If this were not the case, .there would be no point in testing medicines on non-human animals. The way an animal behaves is as much a part of its biology as its hand or the hormones that control its metabolism; in the same way human behaviour, including human aggression and warfare, may also have biological roots.

We can identify behaviour patterns that form part of our biological legacy by looking for those that are common to a number of living primates, and (in particular) are prevalent among our nearest living relatives, the great apes. Comparative studies of living primates show that the facial expressions used during fights are very similar in form and function in different species. In addition, playful fights between peers are a conspicuous feature of the lives of most young primates and most primate species have the capacity to form dominance–subordinance relationships. These behavioural traits seem to be resistant to evolu-tionary change and so may form part of a repertoire of inherited behavioural abilities that we deploy during aggressive encounters.

However, just because we are primates does not necessarily mean that we will behave in exactly the same way as other primates do. Species diverge during evolution, so that while humans may share certain characteristics with non-human primates, in other respects we may well behave differently. This is likely to be the case for patterns of behaviour that are variable among living primates, such as levels of aggressiveness and the social context in which fighting occurs. For such evolutionarily labile traits, which represent recent adaptations to ecological conditions rather than a long evolutionary heritage, we cannot predict how one primate species (including our own) will behave from a knowledge of what even its close relatives do.

In addition, any features of our aggressive behaviour that develop as the result of experience cannot, by definition, have been inherited from our primate ancestors. The pervasive effects of a language-based culture make many aspects of human aggression hard to compare directly with anything other primates do. From birth onwards, children are exposed to a complex set of rewards, punishments, beliefs, traditions and customs, all of which mould their behaviour, including how aggressive they are and the circumstances in which they fight. Language adds its own unique component, being used, for example, both as a weapon and to justify aggressive actions.

Do animals make war?

Certain predispositions that are widespread among living primates, but not those whose distribution is patchy, may underlie part of the behavioural repertoire of our own species. Human warfare is defined as organized, destructive fighting between groups of individuals, using specially designed tools as weapons and with different individuals having specified roles. It takes a very wide variety of forms from small-scale inter-tribal conflicts to global nuclear war. Does anything equivalent occur among non-human animals, and if so, how common is it among our primate relatives?

The vast majority of animal fights take place between pairs of animals and not between groups. In many highly successful species, groups (whether they be schools of minnows, flocks of starlings or herds of wildebeeste) form, merge and break up to the mutual benefit of all concerned and without any aggression at all. There are certainly some species in which social groups defend territories against their neighbours. This group territoriality shares some of the features that characterise human warfare and is sometimes even given the same name. However, examination shows that there is little in the animal world that is directly comparable to human warfare, and nothing that is comparable to modern international warfare. In addition, the phylogenetic distribution of intergroup aggression is patchy, and it is certainly not common among living primates. We cannot, therefore, assume that, just because some animals show intergroup aggression, our own species has inherited this behaviour.

A brief survey of group fighting among non-human animals will demonstrate these points. The most dramatic examples of large-scale, injurious fighting between groups of animals come from the social Hymenoptera, a class of animals (including the ants, bees and wasps) that is very distant from our own species in evolutionary terms. For example, shortly after they emerge in the spring, foraging ants from two neighbouring nests start to fight when they meet, chasing and threatening members of the rival colony and seizing and biting them, dragging them about and stinging them with the specialized ovipositor. Chemicals produced by the fighting ants bring thousands of reinforcement workers from the two nests, resulting in large-scale fighting which may continue for several weeks and cause the death of hundreds of thousands of workers. This is organized destructive fighting by specialised 'disposable' castes using specialized weapons (if not specially made tools) and as such is often called warfare.

Fights between ant colonies are unique, however, and most fighting between groups of animals involves much smaller numbers of animals and is usually less destructive. To give a few examples, a certain species of fish usually lives in breeding pairs within the tentacles of a sea anemone that they defend against other pairs. In certain conditions, where sea anemones are very dense, groups consisting of one breeding female and several, smaller males are found. In such cases, the group members all participate in defence of the territory against intruders. A species of mole rat (a small, hairless and effectively blind rodent) from East Africa lives underground, in extensive burrow systems housing groups of about 75 individuals, dominated by one large adult female (the queen) who does all the breeding. If the colony is invaded by intruders from a neighbouring colony, large, non-breeding mole rats move rapidly to the threatened burrows where they engage in fierce fights with the intruders, some of whom may be killed.

Many carnivores live solitary lives or form pairs. However, some live in groups (and as a result can hunt larger prey) and may combine to defend their hunting area against rival groups. For example, hyaena clans hunt mainly within a defined range from which lone trespassers are chased. Groups of hyaenas patrol their territory, which is marked with strong-smelling anal secretions, and whole clans may come into conflict when one chases and kills prey within their neighbours' territory. In this case, fights may occur and the participants may be injured or (rarely) even killed.

What of primates, whose intergroup interactions are thought to provide the nearest parallels to human tribal fights? Intergroup aggression (though never using specially fabricated weapons) has a patchy distribution but has been observed in several species of Old World monkey and in chimps and gorillas. Within a species, its occurrence is highly variable and dependent on environmental circumstance. To illustrate this point with some well-documented examples, baboons troops usually mingle amicably before separating, with each going its own way. In other species, neighbouring groups avoid each other, and low-intensity aggressive signals and displays (such as the hoots and calls of the gibbon) may be used.

As a general rule, males tend to behave aggressively primarily towards the adult male members of other groups, suggesting that competition for females may be involved. For example, adult male baboons herd their females away from strangers and male vervets show aggression towards immigrants of the same sex and age but may ignore males who enter the range just to feed. On the other hand, aggression by females is often directed towards both males and females, suggesting competition for resources. Female vervets are aggressive towards males and females of all ages, usually over access to food and water. When neighbouring groups of green monkeys encounter each other, the adult males normally display briefly (by jumping about noisily) after which one group may retreat or the resource that has brought them together (often a water supply) may be shared. If the resource is critical, the exchange of signals is longer and more intense, but physical contact is rare and the resource is still often shared. In rhesus monkeys, intergroup relationships are also variable, some displaying high levels of aggression but others involving nothing more than passive displacement. This variability depends, in part at least, on how closely related the members of the opposing groups are.

In gorillas, the social unit consists of one adult silverback male together with a number of adult females and immature animals. Groups occupy extensive, overlapping home ranges with flexible boundaries and usually avoid each other. However, when encounters between groups do occur, these may involve aggressive displays and, though much less often, physical contact and even injury. Common chimpanzees form fluid troops within an extensive home range, with the most frequent form of interaction between groups involving the use of vocalisations to avoid close confrontation. However, in one recent case, groups of adult males have been observed patrolling the edge of the troop range, threatening, attacking and killing members of neighbouring groups. Weapons were never used in these intergroup fights, although chimpanzees do use weapons against predators. Intergroup fighting has never been observed in the pygmy chimpanzee (our nearest relative), although it must be said that this species has been less intensively studied than has the common chimpanzee.

In a small minority of species of non-human animals, therefore, groups do sometimes defend territories and they use coordinated and occasionally injurious fighting to do so. Because this behaviour has such a patchy distribution, there is no reason to believe that ant wars, mole rat wars and human wars have much in common in terms of underlying behavioural mechanisms and certainly no reason to suggest that they have all been inherited from a common ancestor. Non-human primates clearly have the capacity for intergroup aggression, but this is relatively uncommon and does not involve the use of weapons or specialized individual roles. We may have inherited from our primate ancestors the behavioural potential for limited group defence of an area and the resources it contains and this behavioural predisposition may form part of the motivational basis of tribal conflict. However, it is likely to be a very small part and the origins of large-scale modern warfare are historical, sociological, and political rather than biological.

Conclusion

Aggression in animals, and particularly in primates, is a flexible behavioural response, finely tuned to past experience, present conditions and expected future events. As such, it is amenable to control. Humans have probably inherited from their primate ancestors certain simple behavioural predispositions that may be used in aggressive interactions both within and between groups. However, both personal experience and the influences of language-based culture mean that the origins of our more complex personal interactions and social institutions (including modern warfare) are not to be sought in our phylogenetic past.

Further reading

Goodall J. (1986). *The Chimpanzees of Gombe: patterns of behaviour*. Cambridge, Mass.: Harvard University Press.
Goodall, J., Bandora, J. A., Bergman, E., *et al.* (1979). Inter-community interactions in the chimpanzee population of the Gombe National Park. In D. Hamburg &

E. McGowan (eds), *The Great Apes*, vol. 5, pp. 13–53. Menlo Park: Benjamin & Cummings.

Hinde, R. A. (1983). *Primate Social Relationships*. Oxford: Blackwell Scientific Publications. (See particularly chapters by D. L. Cheyney, S. Datta and P. C. Lee.)

Huntingford, F. A. & Turner, A. K. (1987). *Animal Conflict*. London: Chapman & Hall.

Krebs, J. R. & Davies, N. B. (1987). *Introduction to Behavioural Ecology* (second edition). Oxford: Blackwell Scientific Publications.

Maynard-Smith, J. (1984). Game theory and the evolution of behaviour. *Brain Behav. Sci.*, **7**, 95–125.

Wrangham, R. W. (1987). The significance of African apes for reconstructing human social evolution. In W. G. Kinzey (ed.), *The Evolution of Human Behaviour: primate models*, pp. 57–71. State University of New York Press.

4

Is aggression instinctive?

PATRICK BATESON

A pacifist was asked how he would deal with an enemy soldier who attempted to rape his sister. He replied: 'I would interpose my body.' The interviewing tribunal fell about laughing. Nobody could possibly behave like that. Their view was the popular one of human nature. Everybody has an ignition point. Everybody will be driven to attack another person when sufficiently strongly motivated. Furthermore, the motivation to hurt others readily extends from defence to offensive action.

The notion that each of us has a deep instinctive aggressive core prevails to the present day. In part it is based on self-awareness and in part on observations of how humans do in fact behave in special circumstances. However, the popular conception of innate aggression is bundled together with a strange mish-mash of half-truths, false inferences and misconceptions which lead to a view that the animal in us must run rampant from time to time. When this idea is examined critically, however, a rather different view of human nature starts to emerge.

The notion of instinctive aggression is usually coupled with the view that a willingness to harm fellow human beings is a reflection of the beast in man. The aggressive core is part of our biological heritage and 'in our genes'. Since aggression is genetically determined, so the argument goes, it must be inevitable and, in that sense, it is uncontrollable. These thoughts are connected by a thread that is not at all logical, but the *bio*logical link seems obvious enough. So let us start to examine it at the end which firmly clasps the biological theory of evolution.

Evolution of aggression

Virtually every biologist believes that living matter has evolved. Existing species were not created in their present form at the beginning of life on this planet. A lively argument certainly revolves about whether evolution proceeded smoothly or discontinuously. Also biologists quarrel about how variation between individuals was generated and how new species were formed. However, most agree about the evolutionary origins of adaptations, namely those cases where the construction of organisms and their behaviour patterns (when they have them) seem perfectly designed for the jobs the organisms have to perform. Much the most coherent and most universally accepted explanation is that suggested by Charles Darwin.

Darwin proposed a mechanism that depends crucially on two conditions. First, variation in a character must exist at the outset of the evolutionary process which produces a fit between the organism and its environment. Second, offspring must resemble their parents with respect to such a character. The short-term steps in the process involve some individuals surviving or breeding more readily than others. If the ones that survive or breed most easily carry a particular version of the character, that version will be more strongly represented in future generations. If the character enabled them to survive or breed more readily, then the long-term consequence is that the character will bear a close and seemingly well-designed relationship to the conditions in which it worked. If differences between individuals depend on differences in their genes, Darwinian evolution results in changes in the frequencies of genes, although the genes are not, of course, what we see.

Why should aggressive behaviour have evolved? To what kind of problem is it an adaptation? The problem is clearly not one set by physical conditions but rather by the social environment generated by members of the same species. When food (or any other resource which is necessary for life) is in short supply, individuals that can win food will be more likely to survive and reproduce than the losers. Similarly, individuals that are better able to win mates or protect their offspring will be more likely to leave behind progeny (which incidentally are likely to behave in similar fashion). On this basis we should expect aggressive behaviour to occur at all stages of the life-cycle in which individuals may have to compete with fellow members of their own species, but it will be especially prevalent at the time of their lives when they reproduce. This expectation is certainly matched by what is found in humans, but evolutionary theory suggests some complications.

Clearly, risks of injury may be involved when competing for resources or mates or when protecting young. So balances would be expected between avoiding such risks and being aggressive. This is important because each individual may be more likely to survive and reproduce itself if it cooperates with others. Clearly, behaviour that helps another individual can evolve by the means that Darwin proposed. However, the conditions in which it pays to cooperate are often special, and in other sets of conditions, the same individual may be more likely to survive by competing with members of its own species. For instance, if a favour given is unlikely to be returned, then it is better not to give the favour in the first place. In other cases, cooperative behaviour directed towards close relatives is likely to have evolved, whereas identical behaviour directed towards unrelated members of the same species is not. This means that expression of cooperative or aggressive behaviour is likely to be conditional in all those animals that have the equipment to detect changes in conditions. Humans certainly have such equipment.

It may surprise non-biologists that consideration of the biological evolution of aggression should lead to the conclusion that the expression of such behaviour depends on the circumstances confronting the individual. The conclusion immediately invites questions about how aggressive behaviour is controlled and how its development depends on the social environment. So now we should look squarely at the instinct concept and see how it meshes with the notion of conditional expression.

What does instinctive aggression mean?

The term 'innate behaviour' is used widely in biology, psychology and colloquial speech. 'Instinctive behaviour' overlaps extensively with it, but is less commonly used in the technical literature. At least five meanings are attached to innate: present at birth; a behavioural difference caused by a genetic difference; adapted over the course of evolution; developmentally stable; and not learned. Instinct is thought of in very similar terms, but a further and special meaning is also attached to it, namely a distinctly organized system of behaviour driven from within.

The important issue in relation to human aggression is whether such behaviour has all the connotations that whirl around innateness and instinct. Here is a checklist, to which we need answers:

1 Was aggression adapted to its present function by the Darwinian process of evolution?
2 Is aggression genetically heritable?
3 Does aggression appear at a particular stage in development?

4 Does aggressive behaviour have the characteristics of an organized behavioural system?

5 Does aggression develop without previous opportunities for learning?

6 Once present, is the frequency and form of human aggression unchanged by learning?

It is as well to be clear what would be implied if some of the answers were 'No'. When the variety of meanings of 'instinct' are pointed out, some will reply: "We intend *all* of these different meanings to apply to what we regard as a *real* instinct.' I shall call such people the 'pure instinct theorists'. They suppose that there is a part of a Venn diagram where all the areas of different meaning overlap. It has to be said that while the pure position is logically tenable, the instinct concept has been generalized rather loosely, even by the purists, to instances where all the meanings do not apply. The assumption has been, presumably, that if one characteristic is found, the others are implied. That assumption is clearly false in many cases.

A categorical answer to any one of the above questions is liable to trigger a major academic battle. There are immense practical difficulties, for instance, in demonstrating that a piece of behaviour is not learned. Evidence that genetic relatives resemble each other behaviourally is open to the non-trivial objection that they are liable to share common experiences. And so forth. Even so, it is worth examining the shape of the answers which might be given in response to each question.

The characteristics of aggression in humans

*1 Was aggression adapted to its present function by the
Darwinian process of evolution?*

In considering the relevance of evolutionary processes, I have already suggested that they are likely to have played an important role in the case of human aggressiveness, not only influencing the form of the behaviour patterns but also the circumstances in which they are expressed. If this view is correct, then aggressive behaviour was in that sense an adaptation to the conditions in which humans evolved. However, three important qualifications must be made.

First, patterns that were adaptive in the past may not be so any longer. If you fling a nuclear weapon at your enemy, you are unlikely to increase the chances that either you or your family will survive. Second, the development of an adaptive pattern of behaviour in the individual may or may not involve learning. For example, whether or not a person is treated aggressively may depend a lot on how familiar they are. We are clearly prepared to hurt systematically members of an outgroup much more readily than people whom we treat as our own. The ability to recognize close kin is a feature that has almost certainly been shaped by the Darwinian mechanism of evolution and yet kin recognition involves learning. Similarly the inflicting of damage on other individuals may well have been of advantage to those that did it, but the means by which the damage is done is enormously flexible. Uniformity of weapons is not a feature of human aggression. Finally, the expression of aggressive behaviour in the modern world is often controlled by human institutions (such as war) in which individuals adopt particular roles. Such conditions are likely to be quite different from those in which human aggressiveness first evolved.

2 Is aggression genetically heritable?

The most respectable method for analysing the origins of behaviour relies on comparison. In effect, we ask what gives rise to the behavioural differences between individuals. If individuals that are known to differ genetically are reared in identical environments, any systematic difference in their behaviour must ultimately have genetic origins. The logic is impeccable but, as will be seen, the conditions are not as easy to fulfil as they are to state. Moreover, a genetic difference that produces a behavioural difference does not mean that the behaviour pattern in question is unaffected by other factors. If the effects of the genes that matter add together with those of the environmental conditions, then it will be possible to estimate the genetic 'heritability', which tells us how much of the variability found between individuals, when expressing the behaviour pattern, is due to genetic variation. However, if the statistical interactions are non-linear and special combinations of factors produce qualitatively distinct outcomes, then estimates of heritability are meaningless.

Attempts have been made to estimate the heritability of human aggression by comparing identical and non-identical twins. Since genetically identical twins resemble each other more closely on measures of aggressiveness than do non-identical twins, it is argued that genes matter in generating differences. The objection to such a conclusion is that identical twins may have more similar environments as well as being genetically more alike and the method does not disentangle the confounded genetic and environmental sources of variation. A better method is to relate the behaviour of adopted children both to their true parents and to their adopting parents, and compare the correlations. Evidence that children were more like their true parents than their adopting parents would favour the genetic hypothesis. However, this has not yet been done for measures of aggression as far as I know. Another comparision is between the sexes. Men are guilty of far more violent crime than women and the difference has been attributed to the genetic difference between the sexes. Needless to say, the genetic differences are once again confounded with numerous environmental differences.

As things stand, the twin studies have produced heritability estimates of about 0.6, which means that 60 per cent of the variation in the aggression scores is attributed to variation in the genes. The crucial assumption that the influences of the genes and those of the environment *add* together carries simplification to the point of stupidity in my view. However, the evidence does at least somewhat equivocally suggest that genes *influence* aggressive behaviour. That seems to me entirely plausible as a conclusion, even though it in no way implies that humans are inevitably aggressive.

3 Does aggression appear at a particular stage in development?

If we are strict about etymology, innate means 'present at birth' (*in natus*). This descriptive usage is apparently neutral, carrying no implication about developmental processes that occur before birth. The question of where the pattern of behaviour has come from ought to be left open, since opportunities for prenatal learning certainly exist. Even so, 'present at birth' usually does carry a strong implication that the behaviour springs from within the individual. Many people who use 'innate' in this way would also generalize the meaning to include patterns of behaviour which are first expressed at, say, the onset of sexual maturity. The idea behind this broader usage is that the behaviour pattern which is characteristic of the species emerges spontaneously at a particular stage in the life-cycle. In humans it is very obvious that male aggressiveness, in particular, increases sharply at the time of puberty, peaks in early manhood and then declines. This suggests strongly that the factors influencing hormonal state are important in the expression of aggressiveness. Once again, though, beware of facile conclusions. Some of the factors that influence hormonal state are external, such as social stimulation and the quality of diet.

4 Does aggressive behaviour have the characteristics of an organized behavioural system?

Most of the meanings of 'instinctive' are bound up with ideas about biological evolution or about how an individual has developed. However, it is also imbued with a meaning relating to the motivation or short-term control of behaviour. In ethology, 'instinct' was linked to theories about the way behaviour patterns are 'driven from within' by motivational 'energy' which accumulates as long as the response is not being performed. These ideas have been sharply criticized on both logical and empirical grounds and have generally been abandoned as useful ways of explaining how behaviour is controlled in the short term. If a drive to be aggressive exists, it is emphatically not like urine which builds up and has to be discharged from time to time. However, the notion of behavioural systems that are relatively independent of each other has been retained. This usage of 'instinct' relates to the factors that influence the expression of the behaviour. Needless to say, the way in which a system is controlled says nothing about how it got like that. For instance, it might or it might not be influenced by learning.

Aggressive behaviour is clearly organized and is different in its organization from, let us say, feeding behaviour. Threat almost invariably precedes escalation into violence. The emotional state that accompanies the performance of threat and violent acts is affected by conditions that do not affect other aspects of our behaviour. To be caught up in the mood of an angry group is to be swept along in a way which is, for most people, quite different from the surges of feeling they feel in other contexts. All this is real enough, but the fact that aggressive behaviour is organized is irrelevant to an understanding of the conditions in which individuals will behave aggressively.

5 Does aggression develop without previous opportunities for learning?

The issue is whether aggressive behaviour is expressed by people who have not previously experienced the external conditions to which they respond, or without prior practice of the motor patterns which they perform. Unlearned behaviour implies a distinction between experience that has specific influences on behaviour, such as learning, and experience that has general effects. Obviously, food and oxygen, along with a great many other external conditions, are required for normal development of all behaviour patterns. Therefore, the notion of unlearned patterns of behaviour is not applied to activities that develop without experience in the broad sense. No such behaviour could exist. Rather, the idea is that an unlearned pattern develops without the specific experience that could give the activity its particular character. Is human aggressive behaviour like that?

The common usage of 'innate' or 'instinctive' as 'not learned' suffers in practice from the necessity to prove a negative. An enduring classification requires an impossibly large number of observations. The problem is that, having shown that variation in particular environmental conditions does not seem to influence the behaviour pattern, other factors might do so. Alternatively, varying the same set of conditions at a different age might have an effect. In principle, it should be possible to identify unlearned behaviour by systematically excluding likely sources of environmental 'information'. The deprivation (or isolation) condition, which exploits this point, can undoubtedly be of service in eliminating possible explanations. For example, when blind babies smile for the first time about a month after birth, they do not do so because they have previously been studying their mothers' faces. However, apparent common sense must be countered by a number of qualifications:

(*a*) Deprivation can show only what is not learned. It cannot show what is learned during normal development. Excluding possibilities can never show precisely how the behaviour developed.

(*b*) In practice, it is very hard to draw a sharp distinction between experience on which the detailed characteristics of the finished pattern of behaviour might depend and experience which has more global effects on behaviour, in other words, to distinguish between the effects of specific and general aspects of experience.

(*c*) It is often difficult to be certain when an individual will generalise the effects of one kind of experience to what superficially looks like a different context or a different set of activities from those previously associated with the experience.

(*d*) Individuals may have different ways of developing a given behaviour pattern. The conditions of isolation may trigger an alternative way of developing, bringing the individuals' behaviour to the same point as would have been reached if they had experienced the environmental conditions from which they had been isolated. An analogy is provided by commuters who are forced to use bicycles in order to get to work during a fuel shortage. Their ability to get to work does not mean that their cars can run without fuel.

(*e*) Individuals that are isolated from relevant experience in their environment may, nevertheless, do things to themselves that help them to perform an adaptive response later on. By studying yourself in the mirror, you might get a shrewd idea of to whom you are closely related.

After the reservations have been made, though, it is often extremely difficult to understand how learning could possibly be involved in the development of some sequences of movements, such as when a blind human baby smiles. In such cases, the idea of unlearned behaviour is compelling, even though the business of characterizing a behaviour pattern by excluding a role for learning must ultimately rest on plausibility. The first temper tantrums of a child could be quite plausibly treated as being unlearned. The question, though, is what flows from accepting such a conclusion?

6 Once present, is the frequency and form of human aggression unchanged by learning?

Is it the case that patterns of aggressive behaviour, once developed, do not change during the individual's life-time? The answer unequivocally is 'No'. People are like many other animals in that if they are punished for behaving aggressively they are often less likely to be aggressive in the future. Unfortunately for any social policy that might be based on such evidence, they are also likely to become emotionally disturbed, resent the person who administered the punishment, fear the place where it occurred, and become more aggressive in other contexts. On the other side, and once again like many animals, humans are likely to behave more aggressively if they are rewarded for doing so. People are sensitive to the conditions in which they were successful when behaving aggressively, distinguish between individuals with whom such a tactic is likely to work and those with whom it will probably fail, and adapt methods of attack to the weapons at their disposal. All this means that the frequency of aggressive behaviour, the conditions in which it occurs and its form are all affected by learning.

What is more subtle and counter-intuitive is that our attitudes to other people are affected by the way we have treated them. As a consequence, we are often more likely to repeat what we did before. We tend to despise more strongly the individuals towards whom we have behaved aggressively and develop stronger affection towards those to whom we have behaved charitably. Well-designed experiments have shown that, in certain conditions, there is a causal link in what psychologists call 'learning by doing'. A lot more needs to be done in understanding when such effects are obtained. Nevertheless, this line of evidence *does* propose constructive social programmes for decreasing levels of personal violence and for increasing cooperation.

The answer to this final question in the checklist about whether or not aggressive behaviour can be changed by learning, then, is decisively that it can. So the key notion of the pure instinct theorist that human aggression is inevitable and unchangeable is false. This should cause us no serious intellectual problems so long as we identify and abandon the assumption of the pure instinct theorist that behaviour patterns can be divided up into two categories: those which are learned and those which are not learned. So let us take one more look at the behavioural dichotomy which causes so much trouble. Understanding this is crucial to revising the way we think about how our natures develop.

The development of human nature

Why should so many people suppose that our patterns of behaviour can be divided into two types? Why is it supposed that unlearned behaviour patterns are determined by the genes and learned patterns depend solely on particular types of experience? Part of the trouble originates in a common simplification used in science which consists of treating causation as linear. *A* causes *B*, which causes *C* and so on. On this basis, genes cause one kind of behaviour and the processes of learning cause another. The confusion is compounded by the evocative image of a genetic blueprint which clearly implies that a simple relationship can be found between the genetic starting points of developmental processes and their products (unlearned patterns of behaviour). The image is seriously misleading.

I have argued that, like many other patterns of behaviour, aggressive actions depend on internal factors (primarily genes) and external factors (environmental influences). However, like most modern biologists, I assert that no simple correspondence can be found between genes and behaviour patterns. In no sense do the genes represent a scale model of the body. Genes store information coding for the amino acid sequences of polypeptides; they do not code for behaviour patterns. In most cases of complex, long-lived animals, the precise outcome of development cannot be predicted from the starting conditions alone. Development starts with a fertilized egg, comprising a set of active and inactive genes and cytoplasm. These starting conditions create products that change subsequent gene activity, switching some genes off and others on, and exposing the developing animal to new external influences.

It is crucial, then, that we do not confuse the factors influencing development (genes, learning and other forms of experience) with the consequences of development. Both internal and external factors are unquestionably involved in development, but no convincing evidence supports the view that behaviour patterns are of two discrete types, with one group owing its characteristics exclusively to learning and another group owing them to the genes. Nor, as we have seen, is there any reason to suppose that behaviour which is initially unlearned cannot subsequently be modified as a consequence of learning later in an individual's development.

The alternative to a misleadingly over-simplified dichotomy of behaviour is to place the emphasis on the development of individuals as an interplay between them and their environment. The current state of each individual influences which genes are expressed, and also the social and physical world about it. Individuals are then seen as selecting and changing the conditions to which they are exposed. The interplay required for the development of an individual is not between heredity and environment. It is between the individual and environment. This does not mean that the contributions of genetic and environmental sources of variation to behavioural development can never be analysed separately. Nor does it mean that individuals' transactions with their environments modify their patterns of behaviour in all respects throughout their lives. Many examples of behaviour are greatly influenced by certain sorts of experience at one stage of development, in so-called sensitive periods, but are much less affected by similar experience at other stages. Emphasis on the interplay between developing individuals and their environments does mean that we have to abandon thinking about causation as a straight arrow.

If linear ways of thinking are hindering understanding of the origins of aggression, we need some new aids to understanding. One helpful metaphor is the baking of a cake. The flour, eggs, butter and other ingredients react together to form a product that is different from the sum of the parts. The actions of adding ingredients, preparing the mixture and baking all contribute to the final effect. Clearly, it would be absurd to expect anyone to recognize each of the ingredients and each of the actions involved in cooking as separate components in the finished cake. Nor should we expect one-to-one mapping from each word of the recipe to each crumb of the cake. This image is salutary if we are tempted to suppose that behaviour can be dissected into genetic and environmental components.

The baking metaphor leaves out the ways in which developing individuals influence their environments and then are affected by the consequences of their own actions. Another image which does capture those events is a picture of developing individuals involved in games with the social and physical world around them. Each one comes equipped with a set of rules for his or her game, one or more opening moves, and some conditional instructions about what must be done in particular circumstances. Beyond that, nobody can say what is likely to happen in a particular game without knowing the character of the environmental 'opponent'.

The game-playing metaphor has its limitations, of course, but helps us by suggesting that, just as in chess, players help to generate the conditions to which they will later have to respond. Furthermore, their ways of dealing with new conditions will be refined and embellished as they gather experience. The course of a game of chess obviously cannot be foretold even when we know a great deal about the players. By analogy, it is no more possible to predict precisely how a young person will develop in the future from knowledge of what regulates his or her pattern of development. The rules and the person's state influence the course of the developmental game, but they do not exclusively determine it.

Conclusion

To assert that humans are innately aggressive is not totally devoid of meaning, but the misshapen baggage that comes with such an idea is often pernicious and extremely misleading. Humans are most likely to harm others at the stage in life when they are most likely to reproduce. Males are much more likely to injure or kill other humans than are females. Closely related people tend to have more similar aggressive styles than more distantly related people. Much of the patterning and organization appears without being obviously learned. All this fits with a 'biological' view of aggression.

However, modern biology also teaches that the forms of the behaviour patterns that develop in an individual and the frequency with which a given pattern is expressed are heavily dependent on external conditions. That is what is left out so often from the conventional notions of human nature. Even if, perversely, we did not want to bring in cultural and political factors to explain human aggression, we should still expect the large differences that are indeed found between and within human societies. Moreover, we should be aware that our ways of simplifying causality disregard the autocatalytic and feedback character of the way most organisms operate, humans most of all.

The prophecy that, in time, humans are bound to fight each other is liable to be self-fulfilling. Trust and cooperation between people are easily damaged by conflict. If they are reduced, further conflict is more likely. So the modern view of human nature, with its complex dynamics, should cause us to redouble our efforts to emphasize the positive side of social relationships and push the cycle the other way. By doing so, we can help to create the conditions in which people are likely to work with each other. If we do that in an informed and constructive fashion, we have a real chance to make the world a safer and better place.

Acknowledgements

I am very grateful to Robert Hinde for his helpful comments on an earlier version of this chapter.

Further reading

Bateson, P. (1985). The recovery of trust. *Medicine and War*, **1**, 169–76.
Bateson, P. (1987). Biological approaches to the study of behavioural development. *International Journal of Behavioral Development*, **10**, 1–22.
Huntingford, F. & Turner, A. (1987). *Animal Conflict*. London: Chapman & Hall.
Zahn-Waxler, C., Cummings, E. M. & Iannotti, R. (eds) (1986). *Altruism and Aggression*. Cambridge: Cambridge University Press.

5

The genetic bases of aggression

AUBREY MANNING

Is human aggression genetically determined? When the question is posed this way, we must recognize that we may be using the word 'determined' in two senses. First, we are asking whether the expression of aggression can be shown to be due to the action of genes operating during our development. Secondly, we are often asking also whether the expression of our aggressiveness is inevitable, whether as part of our biological make-up it is impossible to evade.

The first question is the more complex and forces us to face up to some formidable problems in its analysis. They are made the more difficult because behaviour as we see it expressed overtly is so remote from the way we know that genetic factors operate. They determine the structure of the proteins we build in the cells of our body but between the proteins and our behaviour there is a large range of developmental steps each of which may have its effect. This makes behaviour genetics a difficult, though fascinating field of endeavour. Hay's (1985) introductory text provides an excellent survey of both animal and human studies.

Genetic analysis of any kind involves breeding experiments and so we must inevitably fall back upon evidence from work with animals. There are many animal species in which all normal individuals, most especially males, show aggressive behaviour. On the other hand, classical genetics relies on the measurement and analysis of the inheritance of *differences* between individuals. A simple morphological example will make the point. People differ in the colour of their eyes and the pattern of inheritance shown means that the difference between having blue eye colour and brown is clearly due to a single gene. Other finer gradations of eye colour involve other genes, but we can say quite unequivocally that blue versus brown is genetically determined, only the genes you carry affect this trait. If we move from the genetics of eye colour to the genetics of eyes *per se*, there is no such clarity. Everybody develops eyes and the vanishingly rare exceptions can tell us little because they appear to be due to gross genetic anomalies. Since there is no variation for the trait, no amount of comparing between relatives will yield any information about genetic control. Of course, there must be a large number of genes which contribute to the complex developmental pathways that lead to an eye. We can also recognize that the environment in which the cells which go to make up the eye grow will have to be adequate for development to proceed. We know, for example, that the position cells occupy in the embryo influences their future fate. The lens of the eye develops under the influence of the eye cup which induces lens formation by those cells which come to overlie it.

Returning to behaviour, many of the same considerations apply. With many vertebrate species, all males normally show aggressive behaviour of their own characteristic type, and we must feel convinced that they all carry some genes which enable this potential to develop. We know next to nothing of how such genes may be operating, far less than we can deduce from the kind of developmental rules which have been discovered for the eye as mentioned above.

This means that our assumption that aggression has a genetic basis is based merely upon commonsense deduction rather than upon genetic analysis. We must remember, too, that where behaviour is concerned we shall certainly not always be safe in assuming that if some pattern is shown by all members of a species, then it must be genetically determined. All chaffinches sing a song which has a highly characteristic 'flourish' at its end. Flourishes certainly vary a lot in detail, but we can recognize that they are always there. Studies of song development show that singing in this way is certainly *not* genetically determined, however; it is copied as a result of hearing adults sing and isolated birds sing songs which lack this most typical chaffinch feature. There are many other examples from animals where elements of their behaviour, especially those characteristic of local populations, can be shown to be acquired in an analogous way to the chaffinch song's embellishment.

Our assumption of an underlying genetic basis for the development of aggressive potential, however reasonable, is not based on hard evidence. If we frame our question concerning the genetic determination of aggression in a different way and ask whether there are genetic factors which affect aggression, then we can move on to much firmer ground. As in all other behavioural features, animals vary in the degree to which they express aggressive behaviour and it can be shown clearly that some of this variation is due to the genes they carry.

Mice are very common subjects for studies of genetics because they breed rapidly and are cheap. Consequently, we know a great deal about their genetic variation, and we have access to various highly inbred lines of mice, the result of brother/sister matings for hundreds of generations. Not all lines survive such inbreeding but a good number of viable lines have been established, and from each of them we can generate animals which are all as similar to each other as are identical twins. Some of these inbred lines are very well known behaviourally. Thus, one famous line, *C57 Black*, is characteristically very active, explores new areas boldly, shows high levels of sexual activity when mated in cages but is only moderately aggressive. In contrast, the *CBA* strain appears nervous and timid in new situations, moving around much less. Its sexual activity is less than *C57* but it is much more aggressive.

All mice tend to fight in the same way; they use the same aggressive and defensive postures, the same manner of biting and so on, but as just mentioned, they vary a good deal in how readily such behaviour can be elicited. Aggression in mice can be measured in a number of ways. Most of them involve matching two males and scoring the latency to the first attack and sometimes how many threats, attacks, bites, etc., are delivered in a standard time, say 5 or 10 minutes. One can also score how many fights are initiated out of 10 or 20 of such trials. It is widely accepted that it is best to measure the behaviour of only one male of the two and to match him against some kind of 'standard opponent', usually a non-aggressive animal who will not fight back. Provided one keeps rearing and testing conditions constant, *C57* and *CBA* score very differently on such tests and it is easy to show that these differences are genetically determined. The F_1 (first generation) hybrids between the two strains show an intermediate score and subsequent F_2 and backcross generations have levels of aggression which indicate that quite a large number of genes are operating.

In comparing inbred strains, we are observing the effects of genetic changes which have accumulated more or less by chance and become fixed, or genetically homozygous, through inbreeding. It is also possible to take an outbred population of animals and selectively breed them solely on the basis of their aggressive performance. Lagerspetz & Lagerspetz (1971) did this with mice and obtained a rapid response to selection, especially in the direction of high aggressiveness. This response was obtained even though it was not possible to measure the aggression score of female mice and breeding was from the sisters of males which had high or low scores. After 11 generations of selection, males of the high line had a mean aggression score of 6.0 (on a 7 point scale), those of the low line 3.1. This result means that in ordinary populations of mice there are a considerable number of genes which affect the level at which aggression is expressed. Natural selection maintains an intermediate level of aggressive behaviour which can be shifted through artificial selection. Similar experiments have been made with other animals as diverse as sticklebacks (Bakker, 1985) and chickens (Guhl, Craig & Mueller, 1960). In a less systematic way, the domestication of dogs and chickens has led to the isolation of aggressive and non-aggressive breeds. The game breed of chickens is fantastically aggressive and this propensity forms the basis of the so-called sport of cock-fighting. Scott & Fuller (1965) in their work with dogs have analysed the familiar differences between some of the terrier breeds which show high levels of aggression and the more placid spaniels and beagles. Again, these differences can be shown to be due to many genes.

In summary, the biological evidence suggests that all animals which show aggressive behaviour carry a number of genes which modify its level of expression. It would be very surprising if human beings were different. It is often difficult to specify how such genes produce their effects because the expression of behaviour can be influenced in so many ways. Genes might affect the amount of the male hormone testosterone which was secreted (testosterone is known to affect aggressiveness in many animals) or change the responsiveness of the central nervous system to the hormone. They might less specifically affect the general level of arousal, or the sensitivity of a sense organ or alter the threshold to pain; any of these changes could also affect the level of aggression shown under certain types of testing. More extensive behavioural testing of selected lines or inbred lines which differ in their aggressiveness might give some clues as to how genes are acting. Lagerspetz found that the mice of her high-aggression line were more active generally and responded more emotionally in other types of test. The problem is to sort out cause and effect from such associations of characters.

It is not usually possible to analyse the way any one gene acts on aggression unless it has a major effect which would enable one to distinguish it from the many other genes which are also operating. We do not have any such examples, but there is one particular genetic element, though not a gene, which has been associated with aggression and this is the Y chromosome. The Y chromosome in mammals is carried only by males and is transmitted from father to son. Thus, if we hybridize two inbred lines of mice, the direction of the cross, i.e. which strain provides the male parent and which the female, will determine which type of Y chromosome the males carry. There will, of course, be other differences between the two types of cross, both genetic – due to the X chromosome – and due to experience, because mothers of the two strains may rear their offspring in different ways. However, by making further crosses of the right type it is possible to breed males who have effectively all their chromosomes in common except the Y. Sometimes such detailed work has revealed that some of the difference between the aggressive behaviour of two strains is due to their Y chromosome. Nearly always the effect is quite small, smaller than the sum of the effects of genes carried on the other chromosomes, but it is nevertheless detectable (e.g. Maxson, Ginsborg & Trattner, 1979).

Such results have aroused considerable interest because hitherto there has been very little information about genes carried on the Y chromosome. The Y is recognized as male-determining, switching the development of the fertilized egg on to masculine lines, but very few genes have been localized on it. Since testosterone is implicated in the control of aggression, there have been various attempts to see whether the Y chromosomes from different mouse strains affect testosterone secretion but none of them has produced convincing results and we have little idea how the Y effect is mediated. However, since aggressiveness is recognized as a masculine quality, there was particular interest in the role of the Y. Furthermore, at about the same time as the mouse studies were undertaken, it became known that there were a few human males with two Y chromosomes (i.e. of XYY constitution rather than the normal XY). Early reports suggested that such men were abnormally tall and abnormally aggressive; in the tabloid press the epithet 'supermale' became inevitable.

Note that the actual link between the mouse and the human findings is very tenuous because with the former we are dealing with differences between Y chromosomes acting in males which have the normal XY constitution, and we can say nothing about the effects of doubling up the Y chromosomes from this mouse work. The human XYY situation is worth discussing because in many ways it provides an object lesson in the pitfalls which await those who rush after simple answers to questions concerning genetic determinism. The second question we asked at the outset of this paper concerned the inevitability of expression of genetic factors for aggression. The XYY story has something to teach us here also.

The first study of a number of XYY men came from Jacobs and her co-workers in 1965. Their paper was entitled 'Aggressive behaviour, mental subnormality and the XYY male'. They surveyed a population of 197 men in a maximum security hospital for mentally disturbed or subnormal men with criminal propensities. Eight – over 3 per cent – had an additional Y chromosome, and this was an astonishingly high proportion for what was then thought to be an extremely rare chromosomal abnormality. Jacobs herself reports that out of 1500 other males whose chromosomes had been checked at her unit for a variety of reasons, only 1 proved to be XYY. Soon after this study a number of others appeared, mostly based upon the inmates of various institutions for dangerous offenders, behaviour disorders or mental subnormality. This may seem an odd way to begin, but during the 1960s there was intense interest in human chromosome abnormalities. The new techniques for studying chromosomes (of which Jacobs herself was a pioneer) had begun to cast light on a number of hitherto mysterious types of mental and physical abnormality. Down's syndrome, one of the commonest forms of congenital disorder, had recently been revealed to be associated with three copies of one small chromosome (number 21) instead of the normal two. It was understandable that the human geneticists followed up this discovery with further examination of mentally retarded patients and, indeed, they did find a high incidence of other chromosomal abnormalities. There is also the mundane but not insignificant point, that the inmates of institutions are a 'captive audience' and it is generally much easier to persuade them to give the necessary blood samples than it is to get samples from the 'normal' population.

Given this kind of selective attention, it is scarcely surprising that an emphasis came to be laid on the hyperaggressiveness of XYY men and this, coupled with their unusual physique, maintained the 'supermale' image. This image was spread by the popular press as events lent themselves to sensational reporting. A mass murderer in Chicago was reported to be XYY; it subsequently turned out that he was not, but the damage was done. In 1968 the defendant in a murder trial offered the fact that he was XYY as a mitigating factor. The response of the court in this case was instructive. Quite reasonably, they rejected the plea after calling for evidence on the incidence of XYY males in the whole population. Jacobs' group had recorded only 1 such out of 1500 'ordinary' men, but gradually the number of people tested has grown, and we now have data from various unbiassed samples of the population at large. It appears that about 0.1 per cent or 1 male in 1000 is XYY, making this, in fact, one of the commoner chromosome abnormalities. Thus, in Britain there will be about 25 000 such people.

However unsatisfactory and incomplete the evidence from institutions may be, every study agrees that the incidence of XYY men among the inmates is considerably higher, sometimes 20 or 30 times higher, than we would expect by chance. Does this mean that XYY men are genetically 'driven' to be hyperaggressive?

It requires careful study to give a sensible answer to this question and such answers will not come quickly. The extra chromosome is there from conception. Throughout infancy and childhood the young boy carrying it is exposed to all the intricate parental, familial and cultural influences as he grows up. One immediate problem for objective assessment concerns the attitudes of parents. Suppose XYY infants are more irritable and throw severe temper tantrums. It may be very hard for parents who know that their child is abnormal and have been already exposed to the sensational image of XYY males to react normally themselves. Their anxieties and distress will inevitably feed back to the child, the foundations of a self-fulfilling prophecy are being laid, and a more disturbed childhood lies ahead. One study has succeeded in overcoming this serious difficulty by comparing the development of two matched groups of boys, one normal and one XYY, whose parents were not informed of their sons' normality or abnormality nor of the real purpose of the study. (Clearly, this procedure raises ethical questions, but parents were given every support as problems arose and provision was made to tell parents if this seemed essential.) This survey continues but preliminary findings have been published (Ratcliffe & Paul, 1986) and it offers the best evidence so far available on the effects of the extra Y.

XYY infants are normal in size at birth but are already larger by the age of 2 and, although not giants, one half of them are in the top 10 per cent of the population for height. They do have an increased incidence of severe temper tantrums as young children and they do have more behaviour problems. Their IQ is somewhat lower than that of the controls (98.4 as compared with 115.7 for the control group, matched for social class, used in this study) and about half of them show delayed speech development and reading problems at school, but they seem to do better than average in mathematics.

In summary, we must conclude that there are genetically determined changes to behaviour resulting from the extra Y, but it would be simplistic to identify the root cause as hyperaggressivity. The increased temper tantrums may result from hyper-reactiveness and frustration.

This may continue into school and the tallness of such boys may lead to their looking older than they actually are with unreal expectations being put on them. A very thorough study of Danish XYY adults by Theilgaard (1984) has revealed that they have more unstable feelings and more lability of mood. Although they do not consider themselves to be more impulsive, that is how the outside world would judge them.

XYY men do have a number of handicaps to overcome; not all succeed and because the end result is sometimes antisocial, they, more often than their normal peers, end up in prisons. In Britain, there may be some 130 of them in maximum security prisons but note that there must be some 24 870 who are not. The huge majority of XYY males lead ordinary lives and the childhood problems which they have are not particularly intractable, i.e. they respond to treatment just as do the problems of chromosomally normal children (Ratcliffe & Field, 1982).

The XYY situation is therefore a good illustration that genetic determination of a behavioural potential does not imply inevitability. There is abundant evidence from work with animals which supports this conclusion. The genetically determined differences in aggressiveness between the various strains of mice are often markedly affected by the conditions under which they grow up. Thus, Scott (1975) and others describe how rearing males of any genetic constitution in isolation typically renders them very aggressive while the opposite is true of social rearing. Lagerspetz's high aggression selected line could easily be made very unwilling to fight by making sure they were defeated in their first few encounters. Similarly, the non-aggressive line males could be boosted by allowing them to win. In general, it appears that experience is even more effective in biasing an animal's behaviour towards or away from aggression than are the genes which it carries. Since we know that the role of individual experience becomes increasingly important in those longer-lived animals living in permanent groups such as elephants and our monkey and ape relatives, we should expect such influences to be at their maximum in our own species. This is not to deny genetic influences over our potential to become aggressive – they are certainly there – but merely to place them in perspective.

We ought to be greatly reassured by the resilience of human developmental processes which obviously enables the huge majority of XYY males to grow up without major behavioural problems. Yet there seems to be a deep-rooted hankering after simplistic, almost fatalistic explanations for our behavioural problems both at an individual and at a societal level. The upsurge of interest in human sociobiology which offers evolutionary explanations for many of our cultural norms is the most modern manifestation of an age-old concern. I would suggest that the emphasis on genetic determinacy is completely misplaced. It remains a matter for investigation whether we can identify genetic factors which affect our normal behavioural potential (it will inevitably be extraordinarily difficult to do so in our species) but even if we gain such information, it can tell us little about what our potential *could* be. More importantly, it can tell us absolutely nothing about how we *ought* to behave and organize our society. Even well-intentioned refutations of simplistic sociobiology such as the book by Rose, Kamin & Lewontin *Not in our Genes* (1985) may by their apparent denial of genetically based biasses seem to imply that, if such biasses were found, they would seal our fate.

It is not difficult to understand how, in the early history of the primates and during the emergence of our own species, those with a moderate level of aggressiveness were best adapted to survive and leave descendants. Can we then escape from our history? The evidence from both animal and human studies suggests that there are numerous genetic factors which can affect the expression of aggression which has, itself, a biological basis. That said, all the evidence goes on to indicate that if we wish to reduce the undesirable manifestations of human aggressiveness, we have the means to do so by consistent and sustained changes to the social climate in which our young people grow up. This will not be an easy task, but it will not be a biological barrier that prevents us from succeeding.

References and further reading

Bakker, T. C. M. (1985). Two-way selection for aggression in juvenile, female and male sticklebacks (*Gasterosteus aculeatus* L.), with some notes on hormonal factors. *Behaviour*, **93**, 69–81.

Guhl, A., Craig, J. and Mueller, C. (1960). Selective breeding for aggressiveness in chickens. *Poultry Science*, **39**, 970–80.

Hay, D. A. (1985). *Essentials of Behaviour Genetics*. Oxford: Blackwell Scientific Publications.

Jacobs, P. A., Brunton, M., Melville, M. M., Britain, R. P. & McClemont, W. F. (1965). Aggressive behaviour, mental sub-normality and the XYY male. *Nature (London)*, **208**, 1351–2.

Lagerspetz, K. M. J. & Lagerspetz, K. Y. H. (1971). Changes in the aggressiveness of mice resulting from selective breeding, learning and social isolation. *Scandinavian Journal of Psychology*, **12**, 241–8.

Maxson, S. C., Ginsborg, B. E. & Trattner, A. (1979). Interaction of Y-chromosomal and autosomal gene(s) in the development of intermale aggression in mice. *Behavior Genetics*, **9**, 219–26.

Ratcliffe, S. G. & Field, M. A. S. (1982). Emotional disorders in XYY children: 4 case reports. *Journal of Child Psychology and Psychiatry*, 23, 401–6.

Ratcliffe, S. G. & Paul, N. (1986). Prospective studies on children with sex chromosome aneuploidy. *Birth Defects: Original Article Series*, 73–118.

Rose, S., Kamin, L. J. & Lewontin, R. C. (1985). *Not in our Genes: biology, ideology and human nature*. London: Penguin.

Scott, J. P. (1975). *Aggression* (second edition) Chicago and London: University of Chicago Press.

Scott, J. P. & Fuller, J. L. (1965). *Genetics and the Social Behavior of the Dog*. Chicago: University of Chicago Press.

Theilgaard, A. (1984). A psychological study of the personalities of XYY- and XXY-men. *Acta Psychiatrica Scandinavica*, Supplement No. 315, pp. 1–133.

6

The physiology of aggression

J. HERBERT

In this chapter, I shall argue that aggression has features that are different from other behaviours. These are important because they influence the way aggression is studied and our attempts to understand its physiological control.

Methods of studying aggression in animals.

Students of aggression try to simplify the conditions under which aggression is investigated, so that behaviour can be controlled, and the factors regulating it dissected out. We understand much about the control of sexual behaviour by studying male and female pairs of rats, put together and allowed to mate. Eating (and drinking) has been effectively investigated by depriving animals (and men) of food and/or water, and then observing both what they did when given access to food (water), and the results of changing the nature of the food, the animal's experience of it, the hormones given to the animal, or interfering with parts of the brain thought to be involved in food intake.

By analogy with these methods of study, the tendency of animals to behave in an aggressive manner can be increased by, for example, their being kept in isolation for prolonged periods, or being given mild electric shocks; staged encounters with another animal then often result in fighting. The problem with this approach is that it ignores the single most important feature of aggression, which is that it has no function or purpose in isolation, but occurs as a component of some other behavioural system. Thus, to study it in isolation is to remove the major question of interest: why do some animals resort to aggressive inter-actions some of the time in order to achieve certain biological goals, and what determines the response of the other animals towards which this aggression is addressed? So, although it is justifiable to study sex, and maternal behaviour and eating in isolation (up to a point) because they can occur in isolation and in response to reasonably specific stimuli and circumstances, the same is much less true for aggression. This is one reason why we know so little about the 'causes' of aggression compared with those of other behaviours, and why the classification of aggressive behaviour is still so unsatisfactory.

Hormones and aggressive behaviour

In many (though not all) monkey species, and in some human societies, the incidence of acts classifiable as 'aggressive' is more common in males than females. This is particularly likely in contexts of sexual competition, or in which the social group is being threatened by an outside source.

'Marking' behaviour (using odours from urine, faeces or special glands) is often the method used to demarcate a territory, and serves both to repel competitors (usually male) and to attract sexual partners. Males of most species mark more commonly than females, though there are exceptions (e.g. marmoset monkeys). Castrating males reduces marking behaviour, as well as the presence of odorous substances in urine, etc. This effect is part of the effect of castration on sexual behaviour, since marking and territoriality are maximal during active breeding. Socially dominant animals mark more frequently than subordinates. Giving the male hormone testosterone to subordinates does not increase their marking behaviour, though castration can reduce that of dominant members. The important point is that this aggression-related behaviour has two separable sets of controls – hormones and 'social' factors.

Many polygamous animals show more aggression towards the same than the opposite sex in many contexts. Treating either sex with the appropriate hormones (testosterone in males, oestrogens in females) accentuates this tendency. However, treating both female mice and women with testosterone has increased their aggressiveness, particularly towards other females. Infanticide also is observed more commonly in males than in females of those species in which it occurs (e.g. some monkeys, rodents under crowded conditions); this, too, is increased by testosterone. Does this mean that androgenic, 'male' hormones are aggression-promoting substances? In the context described above, this seems to be the case. In others, it is either not true or testosterone can actually reduce the likelihood of animals showing aggression.

For example, lactating female rats are liable to be aggressive. Unlike non-parturient animals, such females are often most aggressive towards males, rather than other females. Suckling is required, in many species, to potentiate maternal aggressiveness and maternal responsiveness to the young during the post-partum period. Testosterone given to lactating females reduces their aggressiveness towards males. Hormones regulate aggressive behaviour only in a context-specific manner. Change the context, and the nature of the hormonal control changes as well.

Simpler, and more formalized, methods of studying aggressive behaviour have also been used to investigate the role of hormones. Isolation-induced aggression is more easily observed in males than females (at least in rats). It has been difficult to show sex differences in aggression induced by electric shocks; castration has only a slight effect on this behaviour, unlike isolation-induced aggression. Such experiments, far removed as they are from 'real-life' conditions, again show that hormones will have differing roles in promoting aggression according to the context in which the aggression is elicited.

Hormones also alter the likelihood of an animal being a target. Lactating female rats attack castrated males less than intact males. Adult male monkeys will often not harrass other prepubertal males, but the latters' immunity is lost once they begin to secrete hormones from their testes during puberty; they are then often driven out of the group. Castrating adult males does not seem to restore their prepubertal immunity, so some permanent change independent of subsequent hormonal events seems to have been induced by puberty. Testicular hormones are not unique in this respect: ovariectomized female rats are attacked less than intact ones by lactating females. Hormones cause animals to emit stimuli, both physical (e.g. smell) and behavioural, which cause them to be selected as targets by their conspecifics.

There are, therefore, certain contexts in which testosterone increases aggression. Does that mean that, in such conditions, there is a direct connection between 'male' hormones and aggression? The evidence argues firmly against such a simple correlation.

Hormones and sex differences in behaviour

It was once thought that 'male' hormones induced 'male' behaviour and vice versa. In humans, homosexuality was 'treated' with androgens. But injecting the 'wrong' hormone does not induce behaviour typical of the opposite sex. Indeed, testosterone given to monkeys and women increases their heterosexual behaviour. Clearly, there are differences in the substrates on which the hormones act; the two sexes do not respond in the same way to the same hormone.

Hormones are responsible for some of these sex differences by acting on the brain during early life. Female rats given testosterone just after birth show 'male'-type patterns of sexual behaviour when they are given testosterone later in life. If males are castrated early in life, then they show greater readiness to display 'female' behaviour patterns; they also display much less 'male' behaviour even after testosterone treatment. Male foetuses secrete testosterone during a crucial period of their development; if the brain is exposed to this hormone at this time in its history, then some change is induced that causes a particular pattern of behaviour to be predominant when hormones are again experienced much later in life, at the time of puberty. If no testosterone is secreted, as is normally the case during development of females, then, by default, the 'female' pattern of behavioural response to steroid hormones will be seen in adult life.

The same may be true for certain sorts of aggressive behaviour, particularly that seen in sexual contexts. Males show less aggression during sexual activity after prenatal or early post-natal castration, though the effects have been harder to observe than in the case of sexual behaviour. Attempts to show early hormonal effects on aggression in males in other contexts (as, for example, that induced by electric shocks) have proved very inconsistent. Prenatal androgen given to females has been found to increase their aggressiveness towards other females (but not to males) when they are lactating, thus reducing the relatively sex-specific targeting of this behaviour in the context of maternal behaviour. Girls exposed to similar hormones during intra-uterine life are said to show more 'aggressive' play (a feature of normal boys) than usual. There is now good evidence that human male foetuses secrete androgens during intra-uterine life; there is also a marked secretion during the first few months after birth. It is not yet certain that the first episode has an effect on humans similar to that in rodents; the role of the second remains utterly obscure.

Biochemical evidence finally disposes of the idea that there are 'male' and 'female' hormones. The 'male' hormone testosterone is, in fact, converted to 'female' ones (i.e. oestrogens) by the brain before it can be behaviourally active. In both male and female, the same hormone is acting on the brain, and so one cannot speak of sex-typical hormones and try to link them with sex-typical behaviour patterns. Sex differences in the amount and contexts of aggressive behaviour are dependent upon the nature of the brain on which the hormone acts, not upon the hormone. How far all this applies to humans is still under discussion, though the evidence increasingly points to certain features that are common to the species studied in the laboratory and to man.

Other hormones may also play a part in controlling aggression. Hormones from the adrenal cortex (e.g. cortisol) are secreted in response to a variety of stressors or challenges. Giving cortisol has been found to increase aggressiveness in laboratory situations, which suggests that the endocrine response to such stressors might facilitate an aggressive response to it. The secretion of adrenal hormones is itself controlled by a peptide hormone from the pituitary: ACTH. This hormone has been found to decrease aggression, a paradoxical finding which it is not easy to explain, though some have tried to draw a distinction between 'acute' (cortisol) and 'chronic' (ACTH) behavioural responses to stresses.

Hormone-independent aggression

Maternal aggressiveness seems not to be dependent on either gonadal steroid hormones, or on hormones from the pituitary (such as prolactin). Recently, oxytocin, a hormone released during milk ejaculation, has been shown to induce maternal behaviour after being injected into the brain; suckling is important not only for inducing maternal responses towards the young, but also heightened aggressiveness towards males. Oxytocin could also play a part in the rather special type of aggressive behaviour observed under these conditions. It has been suggested that pups act as anxiety-reducing agents on maternal rats, thus increasing their chances of attacking rather than submitting to male intruders. One of the known effects of giving to humans drugs that reduce anxiety (e.g. benzodiazepines such as Valium) is that aggressiveness may be increased.

Thus, control of aggression by stimuli from the environment and by hormones depends upon the biological context in which aggression is displayed.

Aggression and social status

In the social groups in which most species live, aggression or the threat of aggressive interaction is ever-present. The distinction between aggressive behaviour and dominance status is important. The structure of a society affects the access that individuals have to items in short supply (e.g. food) or items which are not equal in value (e.g. mates); that is, members of the group compete in various ways with each other. Direct aggression, or threat of aggression, may be used to determine which individual will win the competition. Another method is that individuals, through a form of social learning, come to know which member is likely to win such an encounter, and this determines their strategy. This results in dominance hierarchies, an overt form of expression of such learnt roles. Animals or people low in social status may not challenge those higher in the scale, presumably because of the attendant 'cost' in terms of injury, etc. It is important not to mistake levels of aggressive behaviour with status; the two often do not go together.

An observer of a monkey troop recognizes the position of each individual by noting which of the other monkeys it threatens or attacks with impunity (those below it in the hierarchy) and to which it defers (those above it). So the direction, and not the amount, of aggressive interaction reflects the dominance structure of the group. Humans have similar, though more complex, hierarchical structures, some of which are formalized (e.g. in the army). Dominant animals (people) control the behaviour of more subordinate ones; there is now evidence that this process involves endocrine and metabolic mechanisms, as well as behavioural ones, in animals low in rank, and that there are metabolic 'costs' to be paid even by those holding high rank.

It is important for dominant males, for example, to try to maximize their reproductive activity. Study of this process shows how aggression and status interact to control an individual's behaviour. The fact that these mechanisms are efficient is shown by field studies in which it is found that high-ranking primates sire more young than lower-ranking ones. Aggression plays a role. Dominant males may attack subordinates who attempt to mate. They may drive other adult males from the group. But females may prefer males that are dominant to more subordinate ones, even if the latter gain access, though subordinates seem inhibited from trying to mate by their rank.

Status and hormones

Even if subordinates do manage to mate, there are mechanisms that reduce their fertility. Subordinate males have lower levels of testosterone than dominants, and may not only be less potent but also less attractive to females. Subordinate females may ovulate less frequently than dominants, and so be relatively or absolutely infertile. Marmosets, which live in family groups, show this very well: only when the young animals leave their natal groups do they become fertile.

Socially dominant men (who are not necessarily the most aggressive) have been reported as having higher testosterone levels than others; and men exposed to aggression (as in war or in persistently subordinate positions in the armed forces) have lower than normal testosterone levels. There have been reports of correlations between levels of both physical and verbal aggression in men and their testosterone levels, but this is not a constant finding.

Increased population density impairs fertility in some species. Ovulation rates decrease, testicular function diminishes, implantation is reduced and abortion (absorption) rates go up as density increases. So, too, does aggressive interaction, as personal space decreases and the food supply diminishes. The relative contribution of all these factors to the marked impairment of reproduction has still not been worked out. The adrenal hormones are stimulated under these conditions, and may play an even more direct role in the physiological consequences of overcrowding than the gonadal system.

Hormonal response to aggression

The adrenal cortical hormones (corticoids) respond to aggressive interaction. Monkeys living in social groups can show wide differences in cortisol levels. Different authors have not agreed about the relation between cortisol levels and status: some find higher levels in subordinates, others in dominants, and yet others find no clear correlation. In groups that are unstable, or in which social ranks are still unsettled, dominant males may have higher levels of cortisol than subordinates. More settled groups may show different patterns. Changes in the daily rhythm of hormone levels in the blood in response to aggression may be just as important as absolute levels. It is beginning to look as if gonadal hormones are particularly sensitive to an animal's rank, whereas cortical hormones reflect aggressive interactions.

The internal part of the adrenals (the medulla) produces a very different set of hormones, noradrenaline and adrenaline. These are the same as those produced at the endings of certain peripheral nerves which form part of the autonomic nervous system. This system helps regulate the internal economy of the body, such as blood pressure, body temperature and so on. Under many conditions, the autonomic nervous system and the adrenal medulla work together. The two medullary hormones may reflect different aspects of aggressive interaction: noradrenaline is secreted after a fight, it is said, particularly following victory, whereas adrenaline secretion is characteristic of defeat. There is considerable clinical interest in the possibility that certain persistent behaviour patterns (including an 'aggressive' life-style) may promote autonomic and adrenal medullary overactivity, and hence precipitate cardiovascular disease (such as high blood pressure and heart attacks).

Recently, the immune system has been shown to respond to stresses such as receiving persistent aggression. Chronic stress reduces the ability to mount an effective immune response to infection or other foreign bodies. Whether this depends on antecedent changes in hormonal activity – many hormones, including corticoids, can impair immune function – is still not clear.

Aggression and the brain

Are there regions or systems of the brain particularly concerned with the expression of aggressive behaviour? Which systems actually generate a behaviour pattern is an interesting neurological problem, though not of special interest to students of aggression. But is the affective or emotional state of 'feeling aggressive' the same irrespective of the context or circumstances in which aggression is induced? If so, there might be a common neural system underlying such a state, however activated. Secondly, are there systems concerned with the analysis and recognition of potentially aggression-inducing stimuli? Since animals in particular physiological states find certain classes of stimuli aggression-provoking (e.g. males vs other males during breeding; lactating females vs males), there may be a part of the brain that is responsible for classifying input as 'aggression-provoking'. Since many aggressive acts are directed towards objects that induce fear, aversion or threat, the part of the brain that is responsible for 'fear' or 'anxiety' may also be involved in aggression.

Defining neural systems

Although the brain functions as a whole, certain parts of it are specialized for particular functions and are called 'systems'. Papez, in a classic paper, defined the 'limbic' system which, he thought, might be particularly concerned with emotions and their expression. The three areas of the limbic system which have been particularly implicated in aggression are the hypothalamus, the septum and the amygdala.

The limbic brain

The hypothalamus lies at the base of the brain (Fig. 6.1) and is known to regulate many basic bodily functions, such as hormone secretion (from the pituitary), autonomic function (e.g. body temperature) and behaviour (including sexual activity, ingestion and aggression). Damage to the anterior part of the hypothalamus reduces aggressive behaviour by male rats but also abolishes sexual behaviour. If testosterone is implanted directly into this part of the hypothalamus in castrated males, sexual, aggressive and territorial behaviours are all reinstated. This part of the brain is evidently a major site for the coordinated

Fig. 6.1. A: The human brain from the left side. The frontal and temporal cortex occupies a large part of the front of the brain. B–B indicates the plane of the section through the brain shown in B. B: Section through the brain to show the position of the various parts of the limbic system described in the text. The areas each side of the septum are the cerebral ventricles, cavities filled with clear fluid (cerebrospinal fluid).

A.

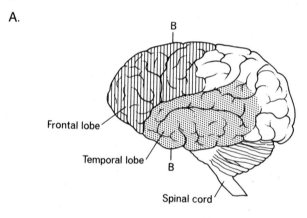

B.

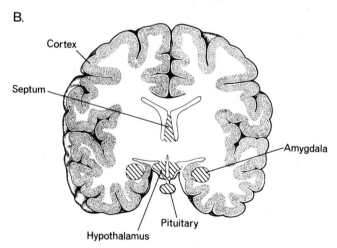

expression of these behaviours. There is some evidence that damage to other areas of the hypothalamus can selectively alter 'offensive' aggression without changing 'defence', so there may be different neural subsystems concerned with functionally different uses of aggression. Electrically stimulating the hypothalamus increases aggression, but stimulated animals attack only 'appropriate' objects (i.e. not more dominant animals), which shows that the direction of aggression is controlled by another part of the brain.

The septum is a small area lying immediately above the hypothalamus (Fig. 6.1), and with important connections with it. Damage to the septum causes animals to over-react to a variety of situations, including aggression-provoking ones. Since other behaviours (e.g. feeding, sexuality) are also altered, the septum is not specifically concerned with aggression, but rather with some wider-ranging function that spans many behavioural categories.

The amygdala is a large clump of nerve cells that lies below the deeper part of the temporal lobes (Fig. 6.1). It is comparatively larger in primates than in carnivores or rodents, etc. Its anatomical connections suggest that large amounts of information enter the amygdala from parts of the cortex, the outer mantle of the brain which performs the most elaborate processing of incoming sensory information. Nerve tracts pass from it to the hypothalamus, and so information is passed to the neural centres concerned with the expression of emotional or 'motivated' behaviour, such as eating, drinking, sex and aggression. Damage to the amygdala alters fear and aggressive behaviour. Animals may lose their fear of novel or unusual objects, and show less aggressive behaviour under those conditions used to elicit it in the laboratory (see above). People who show 'inappropriate' or 'socially disruptive' behaviour have been treated by lesioning the amygdala in Japan and India; the outcome is said to be favourable, though details are often lacking in the published reports. Amygdaloid lesions cause monkeys living in social groups to lose rank, and they may become outcasts. Fear was reduced to novel situations but fear of other (normal) monkeys was actually increased, thus reducing social interaction. Electrical recordings from the amygdala show that its activity increases during social aggression in monkeys, and is reduced by mutual grooming, an activity which seems to have a calming effect. Thus, the amygdala seems to be concerned with socially related aggressive and fearful behaviours. It may be part of the neural mechanisms whereby an animal classifies a stimulus and assesses the relative risk of responding to it, but much work remains to be done before this idea ceases to be speculative.

The front of the cortex, the frontal lobes, overlying the eyes and lying behind the forehead, is much larger in man than any other species, including other primates (Fig. 6.1). Damage to it produces a collection of behavioural abnormalities that include reduced ability to plan actions or to understand timing, and impaired social responsiveness, so that behaviour becomes inappropriate and is not subject to the controls normally exerted by others or by 'social conscience'. There are also emotional disturbances, usually a flattening of emotional responses to events (such as loss, etc.), and direct nervous connections have been found between part of the frontal lobes and the amygdala. The frontal lobes seem to be concerned with some of the more complex functions of social behaviour, including the regulation and use of aggressive behaviour.

Chemical systems in the brain

Detailed studies of the brain's biochemistry show that certain groups of nerve cells use particular chemicals as transmitters, with which they communicate with neighbouring cells. So systems can be identified by the transmitters they use. We now have two ways of looking at the brain: as a series of 'hard-wired' components (the anatomical view) or as a series of chemically related systems (the biochemical view). Each anatomical system may contain several chemical ones, and the same chemical systems may be found in a variety of anatomical positions in the brain, sometimes quite far removed from each other.

Serotonin (otherwise known as 5-hydroxytryptamine or 5HT) is found in many parts of the limbic system in high concentrations. Reducing the levels of serotonin (5HT) increases the display of aggressive behaviour in laboratory animals. Reduced levels have also been reported in humans that commit suicide (self-aggression?) or rape. The interpretation of findings such as these teaches us a great deal about the problems of studying the role of the brain in behaviour. All the serotonin in the brain comes from a few small groups of cells that lie in the brainstem, the part of the brain connecting the spinal cord with the rest of the brain. It is difficult to believe that such a small and widely distributed pathway can carry 'specific' information about any one behaviour. Furthermore, reducing serotonin also stimulates feeding, sexuality and many other behaviours. Therefore, it seems that serotonin has a role in altering the 'response tendencies' or 'reactivity' to a wide variety of stimuli. Of course, this leaves us with the problem of understanding how the serotonin system can be activated or regulated. The serotonin system is clearly very important, but it is not the specific aggression-inducing system for which we are looking, if we believe such a system to be likely.

Peptides are another class of chemical found in the brain. Their interest to students of behaviour is that they are (short) chains of amino acids, and that the structure of this chain is crucial to their activity. This means that peptides themselves contain 'information' much as a string of letters can constitute information (a word). Particular peptides can activate specific patterns of behaviour. For example, one peptide stimulates drinking, another terminates eating, a third facilitates maternal behaviour, and yet another suppresses sexual behaviour. It would be naive to suggest a simple one-to-one correlation between a single behaviour pattern and a peptide, but there does seem to be some 'chemical coding' in the brain which has direct relevance to the control of behaviour. Can we expect an 'aggression-inducing' (or -reducing) peptide? Since aggression, as we have seen, is normally part of other behaviours, it might also be supposed to lack a distinct chemical basis in the brain. But several lines of evidence show that the different components of a more global behaviour can be dissected out in specific parts of the brain, so the display of aggressive behaviour, in whatever context, could be controlled by specific chemicals; that is, there might be an 'aggressivity' peptide. There is already evidence for an 'anxiogenic' peptide, and anxiety can be generated in very different contexts and by very different stimuli. It is possible that particular emotional or affective states may be controlled by particular peptides.

Aggression and brain damage in man

There is no doubt that certain forms of brain disorder can predispose to aggression. Damage to the temporal lobes which, we have seen, contain the amygdala, is known to be associated with uncontrollable outbursts of explosive aggression and rage. This usually occurs as part of a wider 'personality disorder' which includes impulsiveness, undue suspicion and even paranoia. Aggressive outbursts are sudden, extreme, inexplicable and are not followed by remorse. We have already seen that lesions in the same area (which presumably depress rather than excite the surrounding tissue) can cause passivity and, it is said, promote obedience. Encephalitis lethargica is a more generalized infection of the brain. It can have a number of sequelae, which include disinhibited behaviour, emotional lability, reduced 'moral sense' and violent and impulsive behaviour. Unlike some other brain lesions, these people retain insight into their actions. They may also mutilate themselves (self-aggression?) and have to be institutionalized. Other people show episodic 'dyscontrol' which includes aggressive outbursts; they may come from highly adverse social circumstances. The intriguing question is whether these people have brain damage, and, if so, in what way it is related to their social deprivation.

Conclusions

Studies on animals have shown that the various forms of aggressive behaviour are related to particular hormones and chemicals, acting on particular regions of the brain. These systems are also sensitive to social controls and stimuli. Therefore, the display of aggression depends on the way these systems function in the social context in which an animal finds itself. Aggression, like any other behaviour, is never inevitable or wholly predictable; circumstances, constitution and experience make it more or less likely. Everything points to similar principles applying to the control of aggression in man.

Acknowledgements

I am most grateful to Rachel Meller and Jane Rowell for their help with this chapter. The work of my laboratory is supported by Programme and Project Grants from the Medical Research Council.

Further reading

Attili, G. & Hinde, R. A. (1986). Categories of aggression and their motivational heterogeneity. *Ethology and Sociobiology*, **7**, 17–27.

Gray, J. A. (1981). *The Neuropsychology of Anxiety: an enquiry into the functions of the septo-hippocampal system*. Oxford University Press.

Herbert, J. (1987). Neuroendocrine responses to social stress. In A. Grossman (ed.), *Neuroendocrine Responses to Stress*. London: Baillière Tindall.

Hinde, R. A. (1974). *Biological Bases of Human Behaviour*. New York: McGraw-Hill.

Kandel, E. R. & Schwartz, J. H. (eds) (1985). *Principles of Neural Science* (second edition). New York: Elsevier.

Kling, A. & Stecklis, H. D. (1976). A neural substrate for affiliative behaviour in non-human primates. *Brain, Behavior and Evolution*, **13**, 216–38.

Lee, A. K. & McDonald, M. (1985). Stress and population regulation in small mammals. *Oxford Review of Reproductive Biology*, **7**, 261–304.

Lightman, S. L. & Everitt, B. J. (eds) (1986). *Neuroendocrinology*. Oxford: Blackwell Scientific Publications.

Lishman, W. A. (1978). *Organic Psychiatry: the psychological consequences of cerebral disorder*. Oxford: Blackwell Scientific Publications.

Martin, P. (1988). Influences of brain and behaviour on immune function. *Perspectives in Ethology*, **6** (in press).

Schildkraut, J. J. & Kety, S. S. (1967). Biogenic amines and emotion. *Science*, **156**, 21–30.

Selye, H. (1956) *The Stress of Life*. New York: McGraw-Hill.

Svare, B. B. (1983). *Hormones and Aggressive Behavior*. New York: Plenum Press.

Swanson, L. W. & Mogenson, G. J. (1981). Neural mechanisms for the functional coupling of autonomic, endocrine and somatomotor responses in adaptive behavior. *Brain Research Reviews* **3**, 1–34.

C. INDIVIDUAL AGGRESSION AND PROSOCIAL ALTERNATIVES

Editorial

While the previous section was concerned with biological and physiological issues, the present one focusses primarily on psychological ones, specifically the development and elicitation of aggression in the individual. This does not mean that biological factors can be neglected: we have seen that some differences in aggressive behaviour between individuals can be ascribed to genetic differences, and development involves a continuing interaction between the organism at each stage and its environment. Furthermore, the elicitation of aggression may depend on the physiological condition of the individual. Here, however, we are concerned primarily with experiential factors.

Questions of development concern the factors and processes that led to the individual being what he or she is at the time in question. Questions of immediate causation concern the factors and processes that cause an individual to behave in a particular fashion at a particular time, given what he or she is. However, each concerns antecedent events and thus they often merge. For example the 'causes' of mob violence may be sought in the conditions of deprivation in which the individuals grew up, in the recent long hot summer, in the current conditions of unemployment and frustration, and/or in the desire of individuals to outshine their peers in daring. It is usually convenient, however, to distinguish the ontogenetic issues from those of current causation, and to include both situational (or predisposing) and immediately eliciting factors in the latter.

Given the focus of this volume, it would be inappropriate either to attempt a comprehensive review of the literature on the development and elicitation of aggression, or even to list all the factors that have been considered as relevant. But, in general, work on experiential factors in the development of aggression falls into three groups. The first concerns the general social environment. A substantial body of work is concerned with the influence of the mother–child relationship, the familial environment and other intra-family issues. Other workers have been concerned with later social influences, including especially those from peers. Much of this work involves the establishment of correlations between particular types of early environment and subsequent aggressiveness, but the precise mechanisms involved are not the focus of study. For example, several studies show that children brought up in strictly controlling homes lacking in warmth, or children brought up in permissive homes, are more likely to be aggressive than those brought up with firm but reasonable control in a warm and affectionate environment.

The second and third groups concern the mechanisms by which aggressive propensities are acquired or strengthened. One of these stresses reinforcement: aggressive acts are followed by consequences that render their recurrence more likely. The other stresses the role of modelling: the child picks up behavioural styles from influential figures around him. There is little doubt that many other issues await more extensive investigation, for instance, sibling jealousy, or the mechanisms whereby genetically identical twins come to behave more differently if reared together than if reared apart. In any case, none of these influences can be considered independently of the socio-cultural structure: the styles of behaviour that parents reinforce or model depends on their value systems and differs between cultures and even between boys and girls in the same culture. Furthermore, as the following chapters show, it is not profitable to consider the development of aggression independently from that of other types of behaviour, such as prosocial behaviour.

Turning to the immediate causation of aggression, there has been a lengthy and not very useful debate as to whether the causes of aggression should be ascribed to factors internal or external to the individual. Proponents of the former view sometimes implied that endogenously generated aggressive impulses had to have an outlet, those holding the latter view tended to underestimate the importance of the state of the organism. Of course, aggressive behaviour is more likely in some individuals than others, and in any given individual at some times rather than others; and, of course, there are some circumstances in which aggression is more likely than others. The infertility of such either–or distinctions was discussed in the previous section.

The most comprehensive theory of the current causation of aggression ascribes aggressive behaviour to frustration, real or imaginary. This readily embraces the occurrence of instrumental aggression, where the frustration of some form of assertiveness or acquisitiveness leads to aggressive behaviour. It can also be extended to embrace aggression elicited by fear or pain (e.g. frustration of escape), redirected aggression (an individual who is attacked by another dare not reciprocate and redirects his attack on to a third), and even aggression elicited by mere proximity (frustration of the maintenance of individual space). On this view, teasing or hostile aggression must be ascribed to some long-term subjective feelings of being frustrated. The difficulty, it will be apparent, is that the theory of frustration-induced aggression, though fitting many instances, can be too easily extended to explain anything.

In any case, the occurrence of frustration or aggression will depend upon how the actor perceives the situation: whether the frustration is ascribed to another individual or to his/her own shortcomings, whether blame is to be assigned, and so on. Such issues will depend upon situational as well as on the immediately eliciting factors: indeed, whether or not an individual is prone to ascribe blame to self or other is influenced by events in the remote past, and is thus also a developmental issue. The causation of aggression is seldom simple.

These issues are taken up in the chapters in this section. S. Feshbach reviews many aspects of the development and causation of aggression in the individual and L. Berkowitz stresses its situational determinants. N. Feshbach is also concerned with developmental issues, and shows how the balance between aggressive and prosocial propensities can be affected by appropriate training. A. P. Goldstein argues forcefully that the complexity of the bases of aggressive behaviour demands a multi-pronged approach if violence in our society is to be reduced.

Further reading

Bandura, A. (1986). *Social Foundations of Thought and Action: a social cognitive theory*. Englewood Cliffs, New Jersey: Prentice-Hall.

Hartup, W. W. & De Wit, J. (eds) (1978). *The Origins of Aggression*. The Hague: Mouton.

Parke, R. D. & Slaby, R. G. (1983). The development of aggression. In E. M. Hetherington (ed.) *Mussen Handbook of Child Psychology*, vol. 4. New York: Wiley.

Radke-Yarrow, M., Zahn-Waxler, C. & Chapman, M. (1983). Children's prosocial dispositions and behavior. In E. M. Hetherington (ed.), *Mussen Handbook of Child Psychology*, vol. 4. New York: Wiley.

Reykowski, J. (1979). Intrinsic motivation and intrinsic inhibition of aggressive behavior. In S. Feshbach and A. Fraczek (eds.), *Aggression and Behavior change: biological and social processes*. New York: Praeger.

Shaffer, D., Meyer-Bahlburg, H. F. L. & Stokman, C. L. J. (1980). The development of aggression. In M. Rutter (ed.),, *Developmental Psychiatry*. London: Heinemann.

Zillmann, D. (1979). *Hostility and Aggression*. Hillsdale, New Jersey: Erlbaum.

7

The bases and development of individual aggression

SEYMOUR FESHBACH

Introduction

One of the remarkable features of human behavior is its plas-
ticity and variability. We are, indeed, biological creatures who, like other
animals, have to cope with issues of food-seeking, mating and defense.
We too, are engaged in the struggle to survive. But, unlike the insects, we
are not confined by a structure of 'built-in' behaviors dictated by our
genes. We can modify our behavior in accordance with changing
environmental demands, and acquire new behaviors that we also learn to
employ adaptively, selecting different responses for different situations.
And, as humans, we have a rich array of response alternatives from
which to choose. These general properties of human behavior also apply
to acts of aggression. This is not to deny the role of biological factors in
influencing human aggression. But these biological influences do not
operate in a vacuum. Biologically based dispositions to respond
aggressively or non-aggressively can be radically altered by the idiosyncratic experiences of a child and by the behavior and norms of the family
and culture to which he or she is exposed.

Human aggression, then is not a cluster of behaviors that emerge
pre-formed and full-blown in the child or adult. Rather, these aggressive
behaviors are the outcome of a complex developmental process (more
accurately, processes) that have been the object of extensive investigation. In this chapter, we shall attempt to provide a perspective on the
current state of scientific knowledge about the development of human
aggression. One question that immediately comes to the fore is what is
meant by 'aggression'; that is, what are these behaviors or motives or
feelings whose development we are trying to explain? This is not a
pedantic question concerned with academic niceties. The analysis of

what aggression 'is', and of different functions or types of aggression has important social policy implications for the management of aggression as well as theoretical and methodological research implications.

Types of aggression

A wide range of behaviors are often included in the category of aggression. Punching and kicking another child, thrashing of the hands and feet, a derogatory comment, bursting of a balloon, tearing the wing off a butterfly; these are all behaviors that have been labelled aggressive by various investigators. Other forms of aggression include the physical abuse of a child by a parent, the injury to property and person inherent in so many criminal acts, the eruption of rage and destructiveness in a previously conforming adolescent. To these we may also add the violence exerted by the state, at home, in its efforts to maintain conformity to the rule of law, and abroad, in its efforts to pursue its national interest. Further instances of aggression are the destruction of property and manifestations of abusiveness by some student radicals, the more subtle forms of aggression through which men of one color manage to humiliate and degrade men of another color, and, at another level, the violent fantasies sometimes expressed in dreams and in drama. And this enumeration is by no means exhaustive.

It is attractive to utilize concepts such as aggression that embrace a broad range of behaviors and thereby give coherence to otherwise disparate and scattered observations. At the same time, one may be placing under the same rubric behaviors that differ in important ways with regard to their functions and meaning. First, a distinction is required between the emotion of anger and the behavior of aggression. Emotional responses are characterized by heightened physiological arousal reflected in muscular tensions and a particular facial expression. In the case of human anger, one can typically observe a flushed face and either agitated motor movements or a stiffened, tense posture. Violent aggressive acts resulting in pain and destruction do not necessarily accompany anger. There is evidence that anger in adults is relatively infrequently accompanied by aggressive acts and that anger may or may not imply a desire to inflict injury.

As Darwin (1872) recognized, a major function of emotions for animals and humans is one of communication. A friendly display by one organism 'tells' another to approach. An angry display is a signal to an intruder that he or she should retreat or desist. Ethological studies have demonstrated that emotional displays of anger most commonly serve to reduce the likelihood of conflict between two members of the same species; that is, displays of anger by animals are generally not followed by aggressive attacks. Anger can, of course, heighten aggressive behaviors and be associated in humans with a desire to inflict harm. But the connection between anger and inflicting harm is not automatic. To reiterate, the primary adaptive function of anger is expressive, not the infliction of harm; the expression of anger serves as a warning signal to other organisms.

In infrahuman species, there are many instances in which particular stimuli will elicit or 'release' an emotional response, including anger, without the organism ever having previously had direct experience with the stimulus. Most of the situations that make adult humans angry do so as a result of previous experience. While the biological patterning of anger can be observed in human infants, the capacity of particular stimuli and situations to evoke anger in humans is usually learned.

In addition to questions regarding the relationship of aggressive phenomena observed in infrahuman species to human aggression, there are also differences in the interpretation of the significance or function of these aggressive reaction patterns. Konrad Lorenz, a pioneer in the study of ethology and a Nobel Prize winner, defined aggression as 'the fighting instinct in beast and man' (Lorenz, 1966). He conceived of aggression in terms of a biologically based, instinctual system that derives energy from organismic processes independently of external stimuli or provocation. The aggressive 'energy' builds up, as it were, until discharged by an appropriate 'releaser'. In the absence of an appropriate releaser, the aggressive energy will eventually be discharged spontaneously or in response to an inappropriate stimulus.

Lorenz's energy conception of aggression corresponds to Freud's views of human instinctual drives. However, modern psychoanalytic theory has essentially dropped the energy conception of motivation as being theoretically confusing and empirically untestable. In addition, Lorenz's energy-drive conception of aggression has been sharply criticized by ethologists who have questioned the evidence for and necessity of an instinctual energy construct. In summary, there is evidence for biologically based anger and aggressive reaction patterns, and for connections between these reaction patterns and particular eliciting stimuli in infrahuman species. However, there is little empirical basis for assuming a biologically based, innate aggressive drive or aggressive motivation. As the data to be reviewed here indicate, learning and experience play a powerful role in the acquisition of aggressive motivation and aggressive behavior.

In any case, it is often useful to distinguish between instrumental aggression, directed primarily to non-aggressive goals, and aggression with the primary goal of inflicting harm. Shooting another man in defense of one's life or bombing a factory in pursuit of military victory are examples of instrumental aggression. Presumably, if the attacker or the military enemy could be verbally persuaded to desist or modify their provoking behavior, one would use verbal persuasion rather than aggression to achieve these goals. Similarly, a thief who intends to use physical force to steal a wallet will not employ instrumental aggression if the wallet is otherwise accessible by virtue of having been accidentally dropped on the ground. In contrast, the rejected suitor who torments his former mistress and the parent who beats an irritating and disliked child are engaging in aggressive drive–mediated behavior, that is, aggression for the sake of inflicting harm or pain. There are many situations, of course, in which both instrumental and drive aggression are present. Thus, the thief may get some satisfaction from beating his victim, while the parent may gain a more docile child. However, it is useful theoretically and pragmatically, if one wishes to modify the behavior, to determine the extent to which an aggressive act is directed towards non-aggressive goals or towards merely causing harm.

However, in most of the developmental research to be summarized, the focus has been on the overt aggressive response, on the antecedents, manifestation and modification of aggressive behavior, without much attention to the functional nature of the aggressive act. A great deal of important knowledge regarding human aggression can be gained without making operational distinctions between aggressive drive, instrumental aggression and anger. Thus, if a strong link could be shown between consistent use of physical force to discipline a child and the child's subsequent development of violent, anti-social tendencies, those data would be very useful, even if we did not know whether the aggressive child was merely angry or intending to cause harm. The data suggest that modifying the disciplinary practices of parents will result in altering the child's aggressive behavior. However, if we wished to work directly with the aggressive child, then some understanding of the nature of the aggression involved would be very helpful. In addition, if we wish to extrapolate from the developmental findings to such issues as the role of individual aggression in war or the need for society to provide outlets for the expression of aggression, then the specification of type of aggression becomes important.

Theories of aggression

Ethological and psychoanalytic approaches

In the general discussion of types of aggression, we have had occasion to refer to major theoretical approaches to the analysis of aggression. The earlier ethological approach is, in part, reflected in the writings of Lorenz (1966). However, as previously indicated, Lorenz's instinctual energy conceptions of aggressive drive is not representative of the view now held by ethologists. The theoretical conceptions of ethologists vary, but they share a sensitivity to comparative evolutionary issues and a methodology that emphasizes the naturalistic, while systematic and detailed, observation of animal behaviors. An ethologically derived methodology has been successfully applied to the study of dominance behavior in pre-school children, although ethologically based efforts to define aggression topographically in terms of behavior resulting in fight or flight, or 'beating movements' are problematic when applied to human aggression. Parenthetically, it may be noted that despite the enormous body of research on aggressive behavior, there remains a major need for the kind of naturalistic observational study favored by ethologists. For example, the most detailed study available of developmental changes in anger in young children is based on parental reports through diaries, and was published by Goodenough in 1931!

Aggression retains its role as a major human drive for psychoanalytic theorists, although various differences from Freud's conception of aggressive anger, as a biologically rooted form of the death instinct turned outward, have emerged. Psychoanalysts differ in their acceptance of the death instinct and the extent to which aggression is believed to be a reflection of a biological 'need'. They also differ in the role they ascribe to frustration, in the particular frustrations that are assumed to be the primary antecedents of aggression, and in the mechanisms postulated for controlling and experiencing aggressive impulses. Whether or not one agrees with the psychoanalytic interpretations of aggression, there is little question that psychoanalysts have drawn our attention to aggressive phenomena that occur early in the child's development. One of Freud's most significant contributions to the understanding of personality development is his recognition of the deep ambivalence that commonly characterizes the relationships among members of the family unit. Thus, siblings become attached to and love each other. But they also compete for parental attention and favors, and can become envious and aggressive in their competition. The core family conflict for Freud is the rivalry of the child with the same-sex parent. While many psychoanalysts do not accept Freud's libidinal interpretation of the basis for this conflict, generally known as the Oedipal conflict, psychoanalysts of all schools agree on its reality and importance.

A theoretically important feature of these instances of family-based aggression is that the aggressive behavior is linked to particular kinds of frustrating, competitive situations. While psychoanalysts take a pessimistic or tragic view of the human condition in their belief that the child cannot avoid these frustrations, there are, nevertheless, variations in the intensity and frequency of the frustrations and these may then result in variations in children's aggressive behaviors.

As an incidental note, one influential psychoanalytic theorist, Melanie Klein, views the 6-month-old infant as subject to envy of its mother and to intense feelings of hostility. Until detailed observations of infants provide support for these clinically based conjectures, they must remain in the realm of speculation.

The frustration-aggression hypothesis

The psychoanalytic view that aggressive behavior was a consequence of frustration as well as a manifestation of an aggressive instinct provided the basis for the systematic formulation of the frustration-aggression hypothesis by the Yale Learning Theory group (Dollard *et al.*, 1939). The publication of their volume, *Frustration and Aggression*, was a landmark event in the history of social science. Their formulation of the frustration-aggression hypothesis provided a basis for integrating a wide range of aggressive events of interest to sociologists, anthropologists and political scientists, as well as psychologists and biologists. By linking the strength of 'instigation to aggression' to the number and degree of frustrations experienced by the organism, they offered a quantitative formulation that could be investigated empirically. Other hypotheses addressed the effect of inhibition to aggression, e.g. the displacement of aggression to socially acceptable targets.

There is an extensive research literature on children and adults that bears on the frustration-aggression hypothesis. There are studies demonstrating increased aggression in children following a frustrating experience, including evidence of increased social prejudice in a group of boys at camp, following the cancelling of attendance at a movie and the imposition of a series of boring and difficult tasks instead. However there are also a number of studies that have failed to find evidence of increased aggression following frustration, and it is clear that frustration can foster other behaviors besides aggression. Children of 7 to 9 years old can be trained to respond constructively to frustration. In addition, it is evident that the original formulation of the frustration-aggression hypothesis was over-restrictive in its implications that frustration was the only antecedent or cause of aggression. As other theorists have demonstrated, aggressive responses can be learned.

Should the frustration-aggression hypothesis then be discarded as some investigators have suggested? We would argue that the frustration-aggression hypothesis is indispensable for understanding individual differences in children's aggressive behavior. However, the hypothesis needs to be substantially revised and clarified. Only certain kinds of frustration will evoke aggression and only under certain conditions.

First, there are technical questions as to what constitutes a frustration, e.g. should physical pain be considered a frustration? Beyond these technical questions, we know that there are individual differences in children's response to frustration. For example, children who, on the basis of various personality measures, were assessed as having poor self-control responded aggressively when prevented access to a set of attractive toys, while children with strong self-control played constructively with the less attractive toys that were available to them. And we have already seen that children can be trained to respond constructively.

Secondly, which need or motivation of the individual is frustrated makes a difference. For humans, more than for animals, injury to self-esteem may be a particularly powerful elicitor of anger and aggression. Humans are vulnerable to psychological injury, to insult and humiliation, and tend to be more aroused to anger by such injuries than by physical deprivations. One can see the emergence of this pattern in the toddler, during the period popularly known as 'The Terrible Twos'. When the 2-year-old is angered because he or she is denied a request for a toy, or some candy, or to watch a television program, the anger appears to be more a function of interference with the child's emerging autonomy than with the child's desire for the toy, candy, or staying up to watch television. Certainly, by the time the child is a pre-adolescent, frustrations of the need for self-esteem become major determinants of anger and aggression. Parenthetically, this analysis suggests that enhancing a child's sense of security and self-esteem will make that child less vulnerable to psychological indignities and less likely to respond aggressively.

Cognitive factors

The frustration-aggression hypothesis was further modified by the recognition that how we perceive the frustration is a critical determinant of how angry we feel and how aggressively we respond. In the course of daily living, we encounter numerous deprivations, failures and interferences. Whether or not we experience these negative events as frustrating depends to a considerable extent upon our expectancies. If a youngster does not expect to win a race even though he or she would like to, then losing is much less frustrating than if the child had expected to win. Adults who fail to get a salary increment feel very little frustration if no increment was expected, while the same situation could be much more frustrating if a raise had been anticipated. Thus, to understand whether a deprivation will be experienced as frustration and lead to anger, we need to consider the individual's expectancies for satisfaction even more than the absolute amount of deprivation.

There are other cognitive attributes of the frustrating experience that are important to take into account in predicting the likelihood of an aggressive reaction. A frustration is more likely to evoke aggression if it is perceived as *arbitrary* rather than justified. Children are much more likely to be angered by punishment that they believe is unjustified than by punishment that they feel is deserved. A related attribute is the perceived *intentionality* of the frustration. We are much less likely to be angered by the same frustrating or painful event (e.g. being tripped while walking down an aisle) if we perceive it as accidental rather than intentional. This effect has been clearly demonstrated in a number of experiments.

We expect a child to learn the distinction between accidental and intentional frustration and pain. Thus, one of the comprehension items on the Stanford–Binet Test of Intelligence states, 'What's the thing for you to do if a playmate hits you without meaning to do it?' In order to be given credit for the item, the child is required to give a non-aggressive response. By the age of 8, half the children appear to have learned the discrimination.

However, there are systematic individual differences among children in the tendency to attribute hostile intentions to their peers. Most importantly, these differences in attributions are related to individual differences in aggressive tendencies. Highly aggressive boys, ranging in age from 7 to 11 years, are much more likely to interpret an ambiguous provocation of a peer as reflecting hostile intent than their non-aggressive age-mates. The latter react as if the peer acted with a benign intent.

Thus, aggressive boys tend to have a distorted conception of motivations of their peers. They expect hostility from others and are prepared for it. In a kind of self-confirming prophecy, they actually are the recipients of more aggression from others than are non-aggressive boys.

Biological and learning theories

In other sections of this volume, biological antecedents of aggression are considered in some detail. Consequently, we wish only to acknowledge here that genetic and physiological factors have been demonstrated to contribute to differences in children's aggression. In recent years, there has been an increasing amount of research addressed to genetic influences. Greater similarities in ratings of aggression and self-reports of aggression have been reported for identical twins as compared with fraternal twins. A study of adoptees indicated that boys whose biological fathers were criminals were twice as likely to have criminal records as sons whose biological fathers were not criminals.

Current learning theory models of aggression do not reject these biological influences. Rather, they focus on a learning situation that may interact with biological proclivities and, indeed, may override genetically based dispositions. Parenthetically, it should be noted that these genetic studies in humans have not established a specific genetic disposition for aggressiveness. The critical genetic factor might be activity level, emotionality, muscularity or even IQ, with aggression as a secondary consequence.

The learning models have proposed two principal learning mechanisms through which aggression is acquired. One such mechanism is reinforcement of the aggressive response. When a child acts aggressively in order to obtain some goal (another child's bicycle, first in line, approval of peers) and the goal is achieved, then the likelihood of the child behaving aggressively in similar situations is increased. Both experimental and naturalistic studies have demonstrated that through reinforcement, some children learn to become aggressive. Studies of interactions in families with highly aggressive children indicate that the parents often unintentionally reward the child's aggressive behavior. It should also be recognized that parents and other socializing agents may quite consciously reinforce aggression in children. In a study of adolescent boys who had committed delinquent aggressive acts, it was found that their fathers, while discouraging aggression at home, encouraged aggression toward peers.

Studies of sex differences in aggression provide further evidence for the differential reinforcement of aggression. Aggressive behavior is much more likely to be reinforced in boys than in girls. Biological factors also appear to influence sex differences in aggressive behavior. However, despite the fact that differences in aggression constitute the most consistent behavioral difference that has been found between the sexes, these differences cannot be attributed to a simple hormonal or other physiological factor. Administering the 'male' hormone, testosterone, will not, as a rule, produce greater aggression in boys or girls. A number of conditions must be obtained for the hormone to enhance aggression. There is a complex interaction between biological and social factors that results in sex differences in aggression and individual differences in aggression within each sex. Parents who identify physical aggression with 'masculinity' will reinforce aggression in their sons and, as a consequence, will have male offspring who are much more aggressive than sons of parents who do not see physical aggression as a necessary component of masculinity.

Parents and other socializing agents do not have to reinforce their child directly in order for the latter to learn to be aggressive. These socializing agents – parents, teachers, television figures – through their own behaviors, can serve as models. There is a rich body of laboratory findings that indicate that children exposed to models displaying aggressive behavior manifest aggressive responses similar to those displayed by the model. These aggressive behaviors acquired through imitation can persist over some time and are manifested in the absence of the model. By virtue of the process of modeling, children can acquire aggressive behaviors in a rather rapid manner as compared with the more painstaking process of reinforcement. Moreover, as in the case of unintended reinforcement of aggression, models may unintentionally foster aggressive behaviors in the child. Thus, the parent who consistently uses physical punishment to discipline a child for aggressive acts is displaying the very behaviors that are intended to be discouraged. The child then, through modeling, may be learning aggressive behaviors which he or she will then subsequently enact in situations where punishment need not be feared. Thus, modeling may partly account for the finding that one predictor of aggression in preadolescents and adolescents is a history of physical punishment. A related finding is the observation that adults who have been subjected to physical abuse as children frequently become physical abusers themselves.

The parent who uses consistent physical punishment to discipline the child, in addition to serving as an aggressive model, may also be fostering aggression through arousing anger and resentment in the child. These family influences of reinforcement, modeling, and the infliction of pain and frustration may work in conjunction so as to produce a strong and persistent pattern of aggression.

Family influences

Studies of the relationship of features of the family context and of child-rearing to the child's aggressive behavior indicate that the family environment in which the child is reared is an important determinant of the development of aggressive and non-aggressive behavior patterns in children. The child who receives little affection, who is essentially rejected by the family, is then predisposed to the acquisition of aggressive behaviors.

A rejecting environment is undoubtedly a very frustrating environment for the child, thereby fostering aggression. Also, the child does not develop the positive affective bonds that promote prosocial behaviors and inhibit aggression. Another commonly found child-rearing correlate of aggression is permissiveness. The permissive parent may permit aggression to be reinforced by virtue of the lack of discipline or the inconsistent application of discipline. Permissiveness is also likely to be associated with deficient monitoring of the child's activities, and research has indicated that parents of delinquent children engage in less supervision of their child than parents of non-delinquent youngsters.

We have already alluded to the counter-productive character of physical punishment for disciplinary purposes. There are a number of studies that report positive correlations between the amount of physical punishment administered to the child and the development of aggressive behavior. However, the effects of physical punishment are complex, some recent studies finding that aggression is related to the use of only severe physical punishment and others reporting no correlation at all. Interpretation of the correlation between physical punishment and aggression is complicated by the possibility that the aggressive child may elicit more punishment from the parent than the non-aggressive child. Also, low correlations may be obtained if the physical punishment results in the inhibition of overt aggressive acts and in repressed anger and hostility. Thus, studies of extremely assaultive individuals who have been incarcerated for attacks of homicidal intensity indicate that such persons often have a history of mild-mannered overt behavior coupled with buried resentments.

In addition, studies of the likely causal relationships of a combination of early developmental influences on aggression indicate that mothers who are rejecting and negativistic are likely to have more aggressive sons, and that these same mothers also tend to employ physical punishment. Early temperament also appears to be causally related to subsequent aggression. While there are some scholars who feel that the evidence that physical punishment produces aggressive children is ambiguous, one can concur, at the very least, that physical punishment does not foster positive social relationships and behaviors.

Stability of aggression

There has been considerable interest in the question of whether differences in aggression that are manifested early in the child's development, most often assessed during the elementary school period, are also present during adolescence and young adulthood.

The results of the extensive amount of research that has been done on this issue indicate that there is a substantial degree of consistency between early and later manifestations of aggressive behavior. One must be cautious in interpreting these data. These results do not mean that aggression is 'fixed' early in the child's development. A relationship of 0.5 between measures of aggression over a 10-year period is viewed as an impressive relationship, and from a scientific standpoint it is. However, one must realize that even this degree of correlation implies that only 25 per cent of later aggression is accounted for by the measure of early aggression; that is, 75 per cent of the variability in later aggression still remains to be explained. Thus, one can be equally impressed by the changes in aggression that occur over time.

How should one interpret the degree of stability that is found? It may partly reflect genetic and early temperament factors. It may partly reflect early socialization influences, or, more likely, the combination of these biological and experiential variables. In addition, it may reflect the stability of the family atmosphere and social context over the course of the child's development. The development of individual aggressive behavior patterns is best seen as the outcome of a multidimensional set of interacting factors. There are many other factors of importance such as the role of the mass media, and the abundance of aggressive stimuli or cues in the child's environment that we have not discussed here since they are reviewed elsewhere in this volume. And, as other chapters will indicate, one can alter a child's aggressive behaviors through systematic intervention.

References and further reading

Darwin, C. (1872). *The Expression of the Emotions in Man and Animals*. London: John Murray.

Dollard, J., Doob, L. W., Miller, N. E., Mowrer, O. H. & Sears, R. R. (1939). *Frustration and Aggression*. New Haven: Yale University Press.

Goodenough, F. L. (1931). *Anger in Young Children*. Minneapolis: University of Minnesota Press.

Lorenz, K. (1966). *On Aggression*. New York: Harcourt, Brace & World.

8

Situational influences on aggression

LEONARD BERKOWITZ

The issue with which I will be concerned is clear. Where many observers
of human conduct, including Konrad Lorenz and Sigmund Freud, insist
that the violence, destructiveness and cruelty that have been all too prevalent in human history grow mainly out of people's persistent internal
drives, this chapter will try to show that human aggression is largely reactive, a response to situational conditions. It will focus primarily on some of
the research findings highlighting this reactivity, and little will be said
about these conceptions that trace aggression to supposed biological
urges such as a spontaneously generated instinctive drive to violence.
These traditional instinct formulations have been criticized by a good
many writers, and it surely is not necessary to repeat their objections here.
It is sufficient for our present purposes for me only to quote J. P. Scott's
observation published about 30 years ago: 'There is no physiological evidence of any spontaneous stimulation for fighting arising within the body.
This means that there is no need for fighting . . . apart from what happens
in the external environment . . .' (Scott, 1958, p. 62).

Stability and categories of aggression

Saying that a good deal of aggression is largely a response to
situational events does not mean, however, that relatively stable individual characteristics do not play a part in many persons' attacks on
others or that all aggression comes about in exactly the same way. Let us
consider each of these matters.

Persistence of aggressive dispositions

There is considerable evidence that some men have a relatively
persistent predisposition to aggression that can last for years. Studies in

Europe and the United States, making use of behavioral observations in specific situations as well as other people's reports of the person's aggressiveness over a variety of settings, have repeatedly demonstrated that those males who are relatively aggressive as young children are also apt to be highly aggressive as adults. They are also likely to exhibit other antisocial qualities as well, such as fighting, thefts and excessive drinking.

However, these aggressive and antisocial men are not driven to attack others because of pent-up urges inside them that periodically seek a violent release. It is more accurate to say their conduct is the product of some latent qualities that enhance the likelihood of aggressive responses to appropriate situational stimuli. Because of these characteristics, they are inclined to interpret ambiguous occurrences as threats to their pride or as barriers to the attainment of desired goals, become quickly aroused emotionally when they see these threats, and probably are also somewhat unable to restrain their emotional reactions. At least some of these persons are also apt to believe that they can best overcome their difficulties by attacking those who have affronted them or who stand in the way of their getting what they want.

Types of aggression

This last statement identifying some possible motives for aggression brings up yet another consideration. Before we can adequately explain how certain happenings in the external environment give rise to aggressive responses, we must recognize that there are different types of aggression. The kinds of external events influencing one kind of violence are not necessarily the same as the conditions governing other classes of this behavior. Every act of aggression is directed towards the deliberate injury or perhaps even destruction of some target. Nonetheless, this injury is not always the primary goal of the attack. Aggressors sometimes try to hurt or kill their victim in order to attain some other objective such as money or the approval of their peers. In these instances the aggression is instrumental behavior. By contrast, at other times the attacks are carried out chiefly in the hope of doing harm, and any other objectives are secondary to this purpose. Psychologists usually refer to this form of behavior as hostile or angry aggression. Obviously, it is frequently difficult to say whether a particular aggressive act is instrumental or hostile in nature, but it is theoretically possible to make this distinction in general. At any rate, what is important for us here is that some of the factors promoting instrumental aggression do not necessarily heighten hostile (or angry) aggression, and correspondingly, the situational conditions affecting this latter emotionally charged aggression may have a very different impact on the more instrumentally oriented behaviour.

Conditions influencing instrumental aggression

The perceived likelihood of positive outcomes

Since instrumental aggression is carried out in an attempt to reach some goal besides the mere injury of the target, this form of behavior is obviously facilitated by environmental events that (*a*) heighten the salience of this other, non-injurious goal and/or (*b*) strengthen the aggressor's belief that the attack is likely to achieve the desired end. Successful aggression obviously has satisfied many different purposes in human history: wealth, territory, sexual partners, the destruction of opposing belief systems, and the enhancement of a favorable self-identity, among others. But for brevity's sake I will here discuss the effects of situational conditions on instrumental aggression only in terms of people's striving for social approval. To a great extent, this discussion can also be extended to instances of instrumental aggression that are carried out in pursuit of other objectives.

Several prominent sociological accounts of the ethnic and racial group differences in levels of criminal violence, such as the conception of subcultures of violence, basically hold that aggression is often an attempt to win peer approval. The highly violent groups presumably teach their members that they are supposed to act aggressively under certain circumstances and that they can gain status in the group by performing this behavior, or at least they might lose status if they failed to carry out this supposedly appropriate action. In much the same vein, it has also been argued that many instances of 'hooliganism' in the football stadiums of the United Kingdom are actually only instrumental actions intended to bring approval from nearby spectators. Here too, the offenders' peer groups had supposedly taught them that they would enhance their social status by acting in a tough, macho fashion.

Even if we accept these interpretations of the relatively high levels of violence in these particular groups, there is much more to any one violent incident than just learning that aggression can bring social approval. Among other things, there are also the perceived dangers. Although the group members may believe that aggression can be socially beneficial in general, they also know that it is often risky to attack someone; they might be punished in some way for this behavior, by their victim, the police or other persons. The chances that they will exhibit aggression can then be increased by any indications in their immediate environment that this action will yield greater positive than negative outcomes. The sight of others cheering and encouraging aggression conveys such a message. As the other people around them explicitly voice their approval of aggression – or even seem to give this approval by not condemning this behavior when it occurs – it is all too easy for the would-be aggressors to conclude that their aggression will be much more rewarding than costly.

Learning a lesson from observations of violence

The mass media sometimes have just this kind of influence. Parents, educators and concerned citizens have long been troubled about the possible antisocial effects of television and movie depictions of violence, particularly on children. They fear that the frequent TV and movie scenes of people fighting and beating each other up will teach youngsters that aggression is a desirable way of solving their conflicts. Scores of carefully conducted studies have shown that this worry is well-grounded; susceptible viewers can learn what is appropriate conduct in disputes from what they see on the screen. But more than this, the research into media effects also indicates that the portrayals of violence can have unfortunate short-lived influences as well, and on adults as well as on children, even when there isn't long-term learning. Thus, when viewers watch a filmed aggressor benefit from attacking others, many of them come to think, for a short time afterwards, that their own aggression may also pay off. As a result, they could well be more likely to engage in instrumental aggression themselves.

All in all, there is a good chance that susceptible people in the audience will be influenced in some way by the outcome of the portrayed aggression, for a few minutes or perhaps even for a longer period. But however long the effect, we should be concerned about the lessons viewers might draw from the violence portrayed on television and movie screens.

Aversively stimulated hostile aggression

There usually is little doubt about an act of aggression being reactive when it arises in response to some emotional occurrence. What kinds of events are likely to produce these reactions?

Frustrations and aggression

Since at least as far back as the early years of this century, and probably longer ago than this, social scientists have discussed the possible impact of frustrations on aggression. However, as arguments have swirled around the question of the part played by frustrations in this behavior, the participants in the controversy have not always defined their terms in exactly the same way, and as a result, the issues have not always been clearly drawn. I believe we can lessen much of the confusion in this particular area and understand each other better by following the definitions and analysis employed by John Dollard, Leonard Doob, Neal Miller and their colleagues at Yale University in their classic 1939 monograph on the relationship between frustration and aggression.

Frustration as non-fulfilment of an expectancy. Although the Yale group's formulation was couched in the behavioristic terminology of the period, their discussion clearly indicates that they actually viewed frustrations as barriers to the expected attainment of a desired goal. In one place in the book, for example, they quote a poet's question asking,

> Say, what can more our tortured souls annoy
> than to behold, admire, and lose our joy?

Thus, we should not say people are frustrated merely because they do not have something that other persons find attractive. The lack of a color television set is not frustrating to them, for instance, unless they had been hoping and expecting to enjoy the pleasures afforded by this appliance. At any rate, this state of affairs presumably generates an instigation to aggression along with instigations to other activities. The aggressive inclination is not necessarily stronger than the other behavioral tendencies at first, and can be masked by alternative, non-aggression forms of behavior (attempts to solve the problem or to establish substitute goals, for example). However, if these alternative actions fail to overcome the barrier to the desired goal and the thwarting persists, open aggression becomes increasingly more likely.

There is a possible message here for the oppressive dictator who would reduce the amount of violence in his society: eliminate his people's hope of a better life. If they were to expect only minimal satisfactions, they would become resigned to their privations. However, if their hopes should be awakened and then not met, they could become bitterly resentful at the failure to gratify their newly formed desires. Social unrest or even turmoil might then arise.

The frustration need not be 'illegitimate'.   We cannot go into all of the objections that have been leveled against the frustration-aggression hypothesis, but it is worth taking up what is probably the most frequently mentioned criticism. A good many psychologists insist that thwartings provoke aggressive reactions only when they are regarded as having been deliberately produced in an unfair manner. But although this contention is widely accepted, it actually rests on a questionable empirical base. As a matter of fact, several studies indicate that frustrated people can become aggressive even when they cannot attribute their thwarting to someone else's deliberate mistreatment of them. They may be unwilling to attack someone openly because such a reaction could seem inappropriate or even improper; society tells us it is wrong to be nasty to those who had not intended to keep us from our goals. Nevertheless, although the thwarted persons might restrain themselves, they could still be aggressively inclined.

We can see evidence of this frustration-created instigation to aggression in people's responses to competition. Competition can be frustrating. The rivals threaten each other with defeat (which is an anticipatory frustration) and may even actively block the opponent's efforts toward the goal. It is thus not surprising to find in a number of studies that persons competing for victory displayed fairly strong hostility towards their rivals. The competition was legitimate and the others' behavior was entirely proper, but the opponents still became decidedly hostile towards each other.

I will have more to say about all this later, but for now let me offer this concluding thought: although it is widely supposed that people can safely 'discharge' their pent-up aggressive urges in competitive games, the argument I have just summarized, and the evidence on which this thesis rests, indicate that competition can have just the opposite effect. Instead of being a way of reducing persistent aggressive energies, competition might at times generate aggressive reactions.

The target for aggression

After spelling out some of the factors influencing the strength of the frustration-engendered instigation to aggression, Dollard and his associates went on to note that the aggression is most likely to be directed against the perceived obstacle to the goal attainment. However, if the thwarted persons are unable to attack those they hold responsible for their frustration, either because their tormentors are unavailable at the time or because this action is dangerous in some way, they might well displace their aggression and impulsively attack some innocent people. It is important to recognize here that these latter 'scapegoats' for the displaced aggression are often victimized because of their perceived qualities as well as because they are available and safe targets. Not all ethnic minorities are equally apt to be attacked even when they are equally visible and equally unable to retaliate. In various discussions of this topic I have suggested that the displaced aggression is especially likely to be directed against those groups that are associated with the perceived source of the frustration and/or were previously disliked.

The Yale psychologists provided an interesting example of this hostility displacement in social data collected well before the Second World War. Assuming that sudden drops in the market value of cotton were disrupting to the economy of the pre-war American South, since cotton was the major source of income in that region at the time, Hovland & Sears showed that abrupt declines in the value of cotton between 1882 and 1930 were accompanied by significant increases in the number of blacks lynched by whites during that era. The economic frustrations suffered by the whites apparently had evoked aggressive inclinations. The aggression was then directed against those blacks who were strongly disliked, often because they were thought to have violated the South's mores (for example, by seeming to have had sexual relations with white women).

Aversively generated aggression

But let us now return to the competition-produced hostility mentioned earlier. I said that rivalry *can* create hostility, but all of us undoubtedly have been involved in competitive situations in which there were no signs of aggression at all. We can also see this in athletic contests. Athletes rarely attack their opponents even when they are defeated. The theoretical analysis offered by Dollard and his colleagues provides one way out of this apparent difficulty. It contends that the intensity of the overt aggression resulting from a frustration is a function of such factors as the strength of the striving for the goal, how close the person had come to this goal, the availability of alternative responses, the strength of the inhibitions against aggression, and so on. From the Yale group's perspective, then, all of these factors should be considered in predicting whether the rivals in a competition will display open aggression towards each other. However, I would offer a simpler formula: frustrations create aggressive inclinations only to the degree that they are aversive, that is, are decidedly unpleasant to the thwarted persons. Many of the above-mentioned factors (such as how close the thwarted people have come to their goal) presumably determine the intensity of the negative affect produced by the frustration, but basically, it is this negative affect that gives rise to the instigation to aggression.

Physical and psychological pain.   Physical pain is the clearest example of negative affect and a great many experiments have demonstrated that the infliction of pain frequently spurs a wide variety of organisms, humans as well as other species, to attack available targets. This is not to say that aggression is the likeliest response to pain; many animals would rather flee than fight. Genetic background, prior learning and situational influences can all determine what is the preferred response to the aversive stimulus on any one occasion. Nevertheless, the pain activates an instigation to aggression along with instigations to escape/avoid the noxious stimulus, and the aggressive inclination is apt to be revealed in overt attacks if a suitable target is close by, the alternative reactions do not eliminate the aversive occurrence, and restraints against aggression are relatively weak at the time. Keep in mind that the aggression activated here is *hostile* aggression whose primary goal is to inflict injury. The negative affect apparently creates a desire to hurt. The pained animals in one study worked to obtain a target to attack, while in one of my experiments at the human level, young adults feeling severe physical discomfort were more punitive to an innocent bystander when they believed the punishment would harm rather than help this individual.

Psychological as well as physical discomfort can produce the aggression-activating negative affect. The participants in one investigation became hostile after being shown scenes they regarded as disgusting. Also, many of us are hostile towards those who hold attitudes and values greatly different from our own, presumably because the challenge to our attitudes and values is unpleasant. Even depression can arouse aggressive tendencies.

This last statement will not be surprising to many mental health specialists, since the clinical literature abounds with reports of hostility displayed by adult and child depressives. But where psychoanalytic theory holds that the hostility produces the depression – depression supposedly is aggression turned inward – an increasing number of experiments have shown that the experimental inculcation of depressive moods frequently gives rise to angry feelings and even attacks on an available target. At first glance, this observation seems to be inconsistent with the learned helplessness view of depression, holding that depressives are essentially apathetic and often unwilling to exert themselves, but this problem is more apparent than real. Depressives may be reluctant to make an effort, even by aggression, but their aggressive tendencies can frequently be seen in impulsive bursts of temper, especially when they give little thought to what they are doing.

Unpleasant environmental conditions. People can experience negative affect because of the surrounding environment as well as because of their personal qualities or difficulties with others, and there is now good evidence that unpleasant atmospheric conditions often lead to an increase in violence. Thus, research has shown that the urban riots in American cities during the late 1960s were exacerbated by unusual summer heat, and that high temperatures also tend to give rise to 'more normal' violent crimes such as homicides and assaults. Crime statistics can indicate that atmospheric pollution as well as unpleasantly hot days can also contribute to family disorders. Comparable findings have been obtained in laboratory experiments. In these studies, high room temperatures, irritable cigarette smoke and foul odors have all promoted aggressive displays, including relatively strong attacks on an available target.

Associations with unpleasant conditions. Most of us can recognize the validity of these latter observations. We know from our past experiences that we tend to become irritable and easily annoyed when we are not feeling well and/or are in a decidedly unpleasant environment. We might even acknowledge that we are sometimes hostile towards others under these conditions. But we are much less likely to realize that seemingly neutral stimuli in our immediate surroundings can also evoke impulsive aggressive reactions because these stimuli are associated in our minds with aversive events.

Several experiments, with animals and with humans, testify to the aggression-evoking effects of such stimuli. The mere presence of a stimulus associated with strong negative affect can lead to stronger attacks than otherwise would have occurred. In some of the studies dealing with this effect the aversive stimulus was in the surrounding environment. Ulrich & Favell demonstrated that the sound of a buzzer could start animals fighting after this buzzer had been repeatedly paired with painful electric shocks. Fraczek obtained comparable results at the human level. His subjects were much more punitive towards their target when they saw a color that had previously been associated with the receipt of electric shocks.

I have gone on from here to suggest that many persons' reactions to disfigured or handicapped individuals display what is essentially the same kind of phenomenon. While we often sympathize with those who are afflicted or crippled, many of us also associate these people with pain and suffering. The result is that we might be ambivalent toward them. On one hand, we are sorry for them and might want to make them feel better. But at the same time, if we fail to monitor and restrain our actions they are also apt to evoke hostile reactions from us because of their aversive associations in our mind. These unfortunate people, victimized by circumstances beyond their control, can also attract impulsive hostility they do not deserve in their dealings with others.

Conclusion

In so many ways, then, aggression is often a reaction to external events. Whether this behavior is an unemotional attempt to attain some objective other than the injury of the victim or an angry outburst primarily aimed at doing harm, it is incited by something in the surrounding environment and steered by situational cues. The aggressors' internal qualities might well contribute to the attack in important ways: they might have learned in the past that aggression frequently pays off and/or is morally proper in some circumstances. Or these persons might be quick to see dangerous threats and be easily inflamed, perhaps because of prior life experiences but possibly because of their genetic background as well. They may even be 'pre-programmed' by their biological inheritance to become aggressively inclined when they are experiencing negative affect. Whatever the specific determinants of the aggression, people are not innately driven to violence by pent-up forces within them. The violence in our society is best lessened by reducing these external determinants: people must be taught that aggression does not pay and is morally wrong, and that there are alternative ways of responding, other than aggression, when they are exposed to some unpleasant occurrence. The factors producing aggressive reactions can be eliminated, or at least lessened, with some hope of success.

References and further reading

Dollard, J., Doob, L., Miller, N., Mowrer, O. & Sears, R. (1939). *Frustration and Aggression*. New Haven, Conn.: Yale University Press.

Lorenz, K. (1966). *On Aggression*. New York: Harcourt, Brace & World.

Scott, J. P. (1958). *Aggression*. Chicago: University of Chicago Press.

9

Empathy training and prosocial behavior

NORMA D. FESHBACH

Introduction

The concept of empathy and its role in the development of the child raise important theoretical, empirical and applied issues. Empathy, which implies a vicarious sharing between observer and observed, has been hypothesized to have a mediating role in moderating aggressive and fostering prosocial behavior. Empathy, an inferred internal response, does not by itself entail a behavioral transaction such as that reflected by the actual prosocial behaviors of generosity, cooperation or helping. Nevertheless, there are both theoretical issues and empirical factors that link aggressive and prosocial behaviors to each other and each of them to empathy.

Empathy is admittedly an elusive concept, difficult to define, and even more difficult to measure. However, it relates to a facet of human experience and functioning that is intimately involved in our social relationships and social communication. Perhaps human empathy had its origins in the distress behavior displayed by some of our mammalian forebears in response to the distress of another organism to whom they were attached. But early in the developmental process human empathy becomes a more complex phenomenon, involving aspects of the child's cognitive and emotional development. While in some individuals empathy may appear to be an almost automatic, primitive process, for most it is a consequence of learning, socialization experiences and social interaction.

The plan of this chapter is first to review contemporary conceptions of empathy. This will be followed by brief reviews of the relationship of empathy to prosocial behavior and of empathy to aggression. The socialization antecedents of empathy will also be reviewed briefly. Then, the goals, design and findings from an empathy training program developed for children in the middle childhood years will be presented.

The construct of empathy

Systematic investigations of the development of empathy in children are relatively recent. This is surprising, in view of the major role afforded empathy in the social development of the individual and in the psychotherapeutic process (Goldstein & Michaels, 1985). (Given the focus of this chapter, psychotherapeutic issues related to empathy will not be considered directly.) Theoretical conceptions of empathy have varied roughly in parallel with variations in dominant theoretical themes in psychology.

Early formulations of empathy were couched in either predominantly cognitive or affective terms. Later cognitive definitions conceptualized empathy in terms of role taking, perspective taking or social comprehension.

The theoretical debate that occurred in the 1970s, as to whether the nature of the internal response was cognitive or affective, has now subsided and today the general consensus is that empathy entails both affective and cognitive elements, the relative role of each varying with the situation and age and personality of the child. Thus, while empathy is defined as a shared emotional response between observer and stimulus person, it is contingent upon cognitive factors. In the integrative cognitive–affective model I have proposed, the affective empathy reaction is postulated to be a function of three component factors:

(1) the cognitive ability to discriminate affective cues in others,
(2) the more mature cognitive skill involved in assuming the perspective and role of another person, and
(3) emotional responsiveness, that is, the affective ability to experience the other's emotions.

Implicit in this and other models of empathy is the critical requirement for differentiation of self from object.

Hoffman's (1982) developmental model of empathy also has three components – cognitive, affective, and motivational – and focusses on empathic responses to distress in others as the motivation for altruistic behaviors. For Hoffman, empathic arousal is already reflected in infant behavior, and he ascribes reflexive and innate origins to the emergence of empathy. Empathy is also acquired through associative conditioning experiences and through imagination. Within this approach, empathic behavior is primarily affective but subsequently becomes transformed when the cognitive system of the child develops.

The question of the ontogenetic pattern of empathic development is unresolved. If the requirement is that the affective response should reflect cognitive role-taking skills, then it appears that empathic responsiveness emerges during the pre-school period. If an infant's cries in response to a peer crying are considered as empathic behavior, then empathy can be said to emerge almost congenitally. Is this latter behavior 'true' empathy or a precursor to later empathic behavior? Despite the definitional issue and the fact that data in this area are far from complete, a number of generalizations can be made: children at a very early age discriminate emotional signs in others; empathy begins early but becomes more differentiated and purposeful with additional years; not all children are equally responsive or equally empathic.

Empathy and prosocial behavior

How prosocial behavior is defined and measured is related to the theoretical position held. Since the subject of moral behavior has roots in moral philosophy and religion, an additional complexity is added. For the purpose of this analysis, and compatible with perspectives reflected in several recent reviews, prosocial behavior will be broadly defined as behavior that reflects caring or concern for others. In this brief overview no attempt will be made to differentiate between such prosocial acts as sharing, donating, helping, generousness, cooperation and/or altruism.

In general, the data relating empathy to prosocial behavior in adults are reliable. Individuals assessing themselves as being more empathic on paper and pencil self-report measures tend to manifest more helping behavior than individuals who assess themselves as less empathic. In children the relationship is somewhat inconsistent, the direction and significance of the relationship varying with the measure used to assess empathy, the specific prosocial behaviors addressed, the age of the sample and the context of the study.

There are, however, a few research paradigms that do indicate consistency between empathy and prosocial behavior in children. These are studies that focus on the relationship between empathy and cooperation and studies that have attempted to enhance empathic skills through training procedures. Using the Feshbach & Roe Situational Test of Empathy (1968), or a slight modification of it, to assess empathy, a positive association with cooperative play in young children was found in several studies. The Feshbach & Feshbach Empathy Training Study

(Feshbach, 1983), to be described at greater length in the concluding section, also reflected increased prosocial behaviors following training.

The nature of the association between empathy and prosocial behavior in children is not a resolved issue. However a careful reading of the research literature allows a few tentative conclusions. Psychometric limitations of the empathy measures and situational limitations of the prosocial behavior attentuate the relationship between empathy and prosocial actions; the age enhancement of the relationship between empathy and prosocial behavior may be due to the use of self-report measures which may be a more valid, direct index of empathy, or may be a reflection of a more integrated domain in adults of social behavior and cognitive structure. Training empathy skills may be a promising way to encourage prosocial behavior, especially if the training efforts are maintained. Overall, the pattern of findings concerning the relationship of empathy or some component of empathy to various indices of prosocial behavior, such as sharing, cooperation, generosity and helping, has been less consistent than the obtained inverse relationship noted between empathy and such negative behavior as aggression and delinquency.

Empathy and aggression

The inverse relationship between empathy and aggression appears to be a more stable phenomenon, particularly for males. Feshbach & Feshbach's (1969) finding that 6- to 8-year-old boys high in aggression were low in empathy has been supported in a number of different studies employing diverse measures of empathy and aggression and carried out with similar or older age groups. Sometimes a particular component of empathy has been evaluated, such as perspective taking or social sensitivity. In other studies a broader index of empathy was employed. Assessments of aggressiveness included teachers' ratings of aggression, cruelty, and competitiveness and delinquency. One study, contrary to the predominant pattern, reported positive correlations between perspective-taking skills and disruptive difficult classroom behavior in elementary school children.

From a theoretical standpoint, the three-component model of empathic behavior suggests several mechanisms that should result in lower aggression and greater prosocial behavior in the empathic child as compared with the child who manifests little empathy. The ability to discriminate and label the feelings of others is a prerequisite for taking into account the others' needs when responding to social conflicts.

Furthermore, the more advanced cognitive skill entailed in examining a conflict situation from the perspective of another person should result in the reduction of misunderstandings, accompanied by a lessening of conflict and aggression and a greater likelihood of cooperative and other prosocial responses. The assumption of a process of this kind underlies the rationale for the many types of therapy, 'dialogue', and comparable interpersonal communication procedures that have been applied to the resolution of conflict situations.

The affective component of empathy has a special relationship to the regulation of aggression. Aggressive behavior is a social response that has the defining characteristic of inflicting injury on persons or objects, causing pain and distress. The observation of these noxious consequences should elicit distress responses in an empathic observer even if the observer is the instigator of the aggressive act. The painful consequences of an aggressive act through the vicarious response of empathy should function as inhibitors of the instigator's own aggressive tendencies. An important property of empathic inhibition is that it should apply to instrumentally as well as anger-mediated aggressive behavior. Thus, one would predict, on the basis of the affective as well as the cognitive components of empathy, that children high in empathy should manifest less aggression than those low in empathy.

In an extension of this analysis, a comparable affective mechanism has been applied to the analysis of distress. The child who experiences empathically shared distress is motivated to reduce these painful distress feelings. One method for reducing empathically induced distress is to alleviate the distress of the other child through an altruistic, helpful response. Consequently, while the empathic child may find other ways of reducing empathic distress, such as avoidance and denial, one would still expect to find a positive relationship between empathy and prosocial behavior.

Sources of variation in the relations between empathy and behavior

The different relations between empathy/prosocial behavior and empathy/antisocial behavior may be a reflection of the complex structure of empathy. The emotional and cognitive components can vary independently and may affect social interaction and the empathic response in different ways. The emotional component, in particular, appears to be a source of the variability that is frequently observed in this relationship. If the person is emotionally constricted, the empathic response is likely to be flat with less inhibition of aggression and less prosocial behavior. On the other hand, if the person lacks emotional control and differentiation, the empathic response is likely to be excessive and egocentric, obscuring the cognitive aspects of the social interaction and facilitating inappropriate social behavior. Thus, the cognitive component of empathy, a necessary prerequisite for effective social and prosocial behavior, may be rendered ineffective by the intensity of the stimulated affect.

Another possible source of the difference between the relation of empathy to prosocial behavior and its relation to aggressive behavior is the nature of the mediating process. Empathy is presumed to affect aggression through inhibition. No other response is required. However, for social behavior to occur when the child is empathic, the prosocial response must be in the child's repertoire and occur in the eliciting condition.

Still another factor complicating the association between empathy and prosocial behavior is the general issue of the circumstances in which empathy is likely to be evoked. Even highly empathic individuals are not empathic in all situations. Some situations are ambiguous and the affects experienced by the protagonists may be unclear. Or again, there may be conflicting affective and social cues. Still other factors may reduce empathic responsiveness through interference with role taking and perspective taking. For example, it may be difficult to assume the perspective and adopt the framework of an individual whom one intensely dislikes or with whom one is in sharp disagreement or with whom one has very little in common. Yet other situational factors may have such affect-laden significance that they may overstimulate or, conversely, even block affective responses. All these situational contingencies reduce the likelihood of an empathic response and of prosocial behaviors that might be mediated by empathy. Thus, the three-component model of empathy – affective discrimination, perspective taking and affective responsiveness – that provided a basis for examining individual differences in empathy may also suggest a basis for analysing situational sources of variation in empathic responsiveness. Parenthetically, sources of variation suggested by this model can be directly incorporated into the measurement of empathy, and the model can be used as a basis for comparing and analysing different measures.

In summary, empathy and prosocial behavior, although compatible and synchronous, are not automatically or inevitably linked. The inverse relationship hypothesized between empathy and aggression is more consistently obtained. Nevertheless, empathy appears to be one important ingredient of prosocial development. It is a process that facilitates and sustains positive social orientation and responses in a wide variety of conditions and contexts.

The socialization of empathy

Because of the recency of research on the development of empathy, relatively few studies have focussed specifically on the developmental antecedents of differences in empathic responsiveness and behavior. These studies, using different methods and populations reflect, with few exceptions, a consistent picture of the familial contexts that foster or impede the development of empathy. Because of the limited data base on which it rests, this description of the socialization antecedents of empathy should be viewed as a tentative set of generalizations.

The most consistent (and numerous) set of findings are those that bear on physically abusive parents and the relation of a history of physical abuse to empathy in the child. Abusive parents have difficulty in discrimination of emotional cues and manifest lower empathy on self-report questionnaire and projective measures than do non-abusive parents. They display less general empathy and less empathy for children. Moreover, the children of physically abusive parents manifest less empathy than children of non-abusive parents; they obtain lower scores than the latter group on the Feshbach & Roe (1968) Situation Test, a widely used measure of empathy in young children, display less sensitivity to and understanding of the feelings and perspectives of others and much more rarely manifest concern in response to a peer's distress. Being subjected to physical abuse would seem to interfere with the development of empathy.

Turning to more normative populations, the data indicate that parental empathy, as assessed by the Feshbach & Caskey (1985) measure, varies positively with the degree of coherence and absence of conflict in the family and is also positively correlated with the degree of empathy displayed by the child on the Feshbach audiovisual measure of empathy (Feshbach, 1982). The results of a study of child-rearing patterns and the degree of empathy manifested by the child tend further to reinforce this favorable picture of a household that fosters empathy in children. One significant parental correlate of empathy was found for boys – parental emphasis on competition being associated with low empathy. Empathy in girls was negatively correlated with maternal conflict and rejection, with maternal punitiveness and overcontrol, and positively associated with maternal tolerance and permissiveness. In general, empathy in girls is related to a socialization pattern reflecting a positive and non-restrictive relationship with the mother.

Learning to care: the empathy training program

The data indicating that empathy and aggression are inversely related in the elementary school-age child as well as evidence of weaker but nonetheless positive associations between empathy and prosocial behavior led to the design of a training program for elementary school-age children that would enhance the empathic responsiveness of children. The goals of the project were to help regulate aggressive behavior; to promote positive, prosocial behavior; and to develop exercises and strategies that could be used by a regular teacher in the classrooms of children between the ages of 7 and 11 (Feshbach, 1980).

Using the conceptual model for empathy described above, 30 hours of training exercises and activities were developed. The exercises were designed for use in small groups of four to six children. Each activity lasts from 20 to 30 minutes. Efforts were made to include a variety of tasks and to have the materials presented in an interesting and engaging manner. Activities include problem-solving games, story telling, listening to and making tape recordings, simple written exercises, group discussions, and more active tasks such as acting out words, phrases and stories. Several exercises involve videotaping children's enactments and replaying them for discussion. To increase skill in affect identification and discrimination, children were asked to identify the emotions conveyed in photographs of facial expressions, tape recordings of affect-laden conversations, and videotaped pantomimes of emotional situations. In addition, the children themselves role-played in a wide range of games and situations in which they acted out and guessed feelings. To foster children's ability to assume the perspective of another person and to take another's role, training exercises include a variety of games and activities that become progressively more difficult as training proceeds.

Early in the training program, the children were asked to experience and imagine various visual perspectives: 'What would the world look like to you if you were as small as a cat?' They were asked to imagine the preferences and behavior of different kinds of people: 'What birthday present would make each member of your family happiest?' 'What would your teacher, your older brother, your best friend, a policeman do if he found a lost child in a department store?' Children listened to stories, then recounted them from the point of view of each character in the story. Numerous later sessions were devoted to role playing. In these role-playing sessions children were given the opportunity to play a part in a scene, then to switch roles and play the parts of other characters, thus experiencing several perspectives on the same interaction. For other activities children viewed videotapes of their enactments to enable them to gain an outside perspective of themselves and the situations enacted. Discussions followed role-playing sessions and included identification of the feelings experienced by the characters enacted.

A comparable set of 30 hours of training exercises designed for one control group of participants was also developed. These exercises, designed for training in problem solving, focussed on the improvement of students' problem-solving skills in non-social settings. A large proportion of these exercises involved discovery learning through science experiments entailing interesting chemical reactions and work with solutions of different density and solubility. A number of the exercises were games of logic, using attribute blocks and other commercial games. Spatial relations and physical perspective-taking skills were developed through graphic activities and mapping projects. Wherever possible, the control exercises were very similar to the empathy exercises in format, with only the content varying. For instance, if the empathy training groups played the game 'Concentration' and matched facial expressions (happy to happy, angry to angry, etc.) of the various individuals portrayed on the game cards, the control group played 'Concentration' but matched shapes printed on the same card.

Participants in empathy training studies were 8-, 9- and 10-year-olds. However, the exercises, with some minor modifications, can be used with children aged from 6 to 11.

Findings from field studies carried out in the Los Angeles City Unified School District indicated that children who had participated in empathy training showed statistically significant changes on both the aggressive and prosocial dimensions. Also, following the empathy training, children reflected a more positive self concept than children in the problem-solving control condition and children who had received no training. Of note is that children in the problem-solving control activity groups also displayed a decline in aggressive behaviors. Taking these children out of the classroom and providing them with positive small group experiences facilitated a reduction in aggression, whether or not the training experience focussed on empathy or on interesting academic content. However, empathy training proved to be critical in regard to positive shifts in prosocial behaviors. Children in the empathy training condition differed significantly from both control conditions in the increase of incidents of such prosocial behaviors as cooperation, helping and generosity as assessed by teacher reports of critical incidents. In the case of prosocial behaviors, merely taking the children out of the classroom was not sufficient. These behaviors were facilitated by specific training in affect identification and understanding, role playing and emotional expressiveness, components believed to be implicated in the empathic process. It would appear that empathy training helps to bring out more positive social behaviors and a more positive self-evaluation in aggressive and in non-aggressive children.

Conclusion

The data that have been reviewed here indicate that empathy is a meaningful psychological attribute that is associated with lesser degrees of aggression in children and in adults. Empathy is also associated with prosocial behaviors, more consistently for adults than for children. The data further indicate that empathy is linked to significant developmental experiences of the child; that physical abuse and family conflict and lack of cohesion impede empathy in boys and girls, while a non-competitive paternal focus facilitates empathy in boys and a positive accepting maternal attitude that permits independence is conducive to the development of empathy in girls.

In addition to empathy varying with childhood experiences, it also appears that different components of empathy can be enhanced through training. Empathy training exercises derived from the three-component model of empathy can be readily incorporated into the school curriculum. In addition, this model may provide a guide for the enhancement of empathy in other contexts in which we wish to reduce aggressive behavior and foster positive social feelings and interactions.

Perspective- and role-taking exercises, for example, could be designed to entail taking the perspective of other ethnic and national groups and assuming their role in the enactment of current conflict situations. Particular emphasis might be given to the feelings experienced by protagonists from one's own and other groups in enactments of these situations. In addition, there are data indicating that the greater degree of perceived similarity between two individuals, the greater the degree of empathy (Feshbach & Roe, 1968), suggesting that delineating similarities between individuals in a potential conflict is a useful strategy. Finally, from a theoretical standpoint, there is reason to anticipate that responding empathically fosters feelings of attachment toward the empathic individual. The therapeutic situation provides an excellent example of this process. Therapeutic interventions conveying understanding of and responsiveness to the clients' perspective and emotions can elicit positive feelings towards the therapist. Fostering empathy and responding empathically in potential conflict situations may not eliminate the conflict but should contribute to its resolution and reduce the likelihood of an aggressive outcome.

References and further reading

Feshbach, N. D. (1980). *The Psychology of Empathy and the Empathy of Psychology.* Presidential address, presented at the annual meeting of the Western Psychological Association, Honolulu, HI.

Feshbach, N. D. (1982). Sex differences in empathy and social behavior in children. In N. Eisenberg (ed.), *The Development of Prosocial Behavior*, pp. 315–38. New York: Academic Press.

Feshbach, N. D. (1983). Learning to care: a positive approach to child training and discipline. *Journal of Clinical Child Psychology*, **12(3)**, 266–71.

Feshbach, N. D. & Caskey, N. (1985). A new scale for measuring parent empathy and partner empathy: factorial structure, correlates and clinical description.

Feshbach, N. & Feshbach, S. (1969). The relationship between empathy and aggression in two age groups. *Developmental Psychology*, **1** 102–7.

Feshbach, N. D. & Roe, K. (1968). Empathy in six- and seven-year-olds. *Child Development*, **39**, 133–45.

Goldstein, A. P. & Michaels, G. Y. (1985). *Empathy: Development Training, and consequences*. New Jersey: Erlbaum Associates.

Hoffman, M. L. (1982). Developmental prosocial motivation: empathy and guilt. In N. Eisenberg (ed.), *The Development of Prosocial Behavior*, pp. 218–31. New York: Academic Press.

10

Aggression reduction: some vital steps[1]

ARNOLD P. GOLDSTEIN

Aggression in its diverse individual and collective forms has long been a world-wide problem of the first magnitude. When viewed in global perspective, contemporary aggression takes many guises: violence and vandalism by juveniles; child and spouse abuse and other forms of domestic or familial violence; assaults, muggings and homicides; rape and other sex-related crimes; politically motivated terrorism; racially or economically motivated mob violence; and aggression in many forms directly or indirectly initiated by the state. And this is far from an exhaustive list. We could add athletic mayhem, clan blood feuds, ritual torture, police brutality, organized warfare, and much more. The variety, intensity, frequency and overall prevalence of overt aggressive behavior throughout the world is starkly and appallingly high (Goldstein & Segall, 1983, p. vii).

Strategies for aggression reduction

What solutions have been offered towards the goal of reducing or moderating such diverse and substantial expressions of human aggression? Three popular strategies have emerged: control, catharsis and cohabitation. We believe all three to be destined to failure, and wish now to describe briefly these perspectives and the bases for our belief in their inadequacy.

Control

The control philosophy, promulgated most successfully in eras of political conservatism, is clearly exemplified by the thrust of changes

---

[1] This chapter is a somewhat shortened version of a more fully documented article. The latter is available from the author on request.

that have taken place in the response of America's criminal justice system to violent crime. With regard to juvenile offenders, 'the rehabilitative ideals of the juvenile court are being reviewed and overshadowed by concern with community protection, punishment, retribution, and increasingly secure confinement' (Fagan & Hartstone, 1984, pp. 31–2).

Similarly, Bartollas (1985) comments *vis-à-vis* adult offenders (p. 14):

> Proponents of the hard-line approach ... see punishment as more likely to deter crime and to provide protection to society. They also claim that the establishment of law and order demands firm methods of crime control, such as a greater reliance on incapacitation, the use of the death penalty, the implementation of determinate and mandatory sentences throughout the nation ...

But are punitive means intended to control aggression in fact successful? Most relevant research yields negative results (Hagan, 1982). Ross & Fabiano (1985) summarize these findings clearly (p. 164).

> Very low correlations have been found between crime rate and the certainty of apprehension or the severity of punishment. Sometimes the correlations have been negative, i.e., the greater the chances of being caught and punished, the *higher* the crime rate! Deterrent effects are frequently unreliable and weak ... Deterrence effects, when they are achieved, are typically only temporary. Often they serve only to change the kind of crime committed to those which have not been made a focus of attention. Sometimes, they only serve to shift crime to another geographic location.

In the home, and often in the school, control of aggression is often sought through verbal or physical punishment techniques. Their effectiveness has been shown to be a function of several factors:

1 likelihood of punishment;
2 consistency of punishment;
3 immediacy of punishment;
4 duration of punishment;
5 severity of punishment;
6 possibility of escape or avoidance of punishment;
7 availability of alternative routes to goal;

8 level of instigation to aggression;

9 level of reward for aggression;

10 characteristics of the prohibiting agents.

Punishment is more likely to control aggressive behavior the more certain its application, the more consistently and rapidly it is applied, when it is introduced at full intensity rather than gradually increased, the longer and more intense its quality, the less likely it can be avoided, the less available are alternative means to goal satisfaction, the greater the level of instigation or reward for aggression, and the more potent is the prohibiting agent. Given these several contingencies, it is not surprising that most investigators report the main effect of punishment to be a *temporary* suppression of aggressive behavior. Punishment research has also demonstrated such undesirable side-effects of punishment as withdrawal from social contact, counteraggression towards the punisher, modeling of punishing behavior, disruption of social relationships, failure of effects to generalize, selective avoidance (refraining from aggressive behaviors only when under surveillance), and stigmatizing labeling effects.

The evidence is not all one-sided. It has been reported that punishment can lead to rapid and dependable reduction of inappropriate behavior, and the consequent opening up of new sources of positive reinforcement, improved social behavior, discrimination learning, and other positive side-effects. Nevertheless, on balance a broad conclusion seems apparent. As with the expressions of the control strategy in the context of America's criminal justice system, control as the route to reducing aggression in the home or school leaves much to be desired, in both dependability and effectiveness, as well as in its ethical limitations.

Catharsis

An especially popular view of how aggressive behavior is optimally managed promoted catharsis as the solution. Catharsis refers to the draining off or venting of an emotion, an experience which purportedly may occur vicariously, by observing and empathically identifying with another person, or directly via one's own behavior. The concept of catharsis first appeared in the vicarious sense, in connection with the reported emotionally purging experience of the audience at early Greek drama. More specifically related to aggression is the Freudian view that '... there is a continuous welling up of destructive impulses within the individual representing an outgrowth of the death instinct' (1950, p. 160). In such an hydraulic view, therefore, aggression is inevitable and the best one can accomplish is to direct, channel or regulate its periodic release or discharge via a socially acceptable, minimally injurious 'aggressive' act, e.g. debates, the space race or, of direct relevance to several of the examples we will offer below, competitive sports.

The view of aggression as stored, constantly growing energy, which can be directly or vicariously vented from the person's reservoir, is widely held. Is it correct? Research has been abundant and of several types.

Static comparison. One approach is simply to compare the aggressiveness levels of persons who do or do not regularly engage in aggressive activities. If the catharsis notion is correct, those who do, having vented, should show less aggression. Comparisons between contact sport athletes (football, wrestling), non-contact sport athletes (swimming, tennis) and non-athletes have shown no between-group differences on behavioral measures of aggression. Such studies also provided no evidence that expressing aggression reduces its occurrence.

Before-after comparisons. These are studies in which two groups are identified or randomly constituted, and one but not the other is given the opportunity to aggress. If the catharsis concept is correct, the after comparisons should reveal that the former group is less aggressive than the latter. No support for such an effect has been found. Indeed three before–after comparison studies of catharsis found an *opposite* effect. Persons permitted to behave aggressively afterwards became more, rather than less, aggressive. Goldstein & Arms (1971) measured aggression levels in randomly selected fans just before and after an Army–Navy football game. Contrary to catharsis, fans of both winners and losers increased pre- to post-game in their level of aggression. The investigators comment, 'Exposure to the aggression of others seemingly acts to weaken one's internal mechanisms controlling the expression of similar behavior' (p. 165).

Archival studies. These are examinations of short-term or long-term records of aggression-expressing sports and other events which, if catharsis is a real phenomenon, should reveal its decrease over the course of the event. In five separate studies, Russell (e.g. 1983) showed the opposite of a catharsis effect. Since catharsis by definition predicts aggression should decrease as it is expressed, progressively less should occur as an athletic event progresses. In fact, aggression *increased* as the games progressed. A similar anti-catharsis result emerged in the same author's tracking of aggression between two teams the more times they met over a season: more and not less aggression took place. Other studies have shown positive relationships between the degree to which combatant sports existed in a society and its involvement in wars, conflicts, revolutions and similar events, and between whether a country participated in the Olympics and the number of athletes it sent, on the one hand, and both the number of wars in which it participated and its number of months at war on the other.

Laboratory studies.   Fully consistent with these results, Berkowitz (1964) conducted a series of investigations in which research subjects were or were not shown either brutal but staged sports violence (e.g. the fight scene from the movie *The Champion*), or equally brutal actual fights from hockey, football and basketball. In all instances, those viewing the aggressive scenes, as compared with non-viewing control group subjects, showed increased levels of aggression.

The conclusion is clear: catharsis is a myth.

Cohabitation
By cohabitation as a further prevailing perspective on the reduction of aggression, we mean the defeatist, resigned sense that aggression is 'human nature', 'will always be with us' and one has no choice but 'to live with it'. Many holding this view live with it poorly, e.g. the many thousands of city-dwelling elderly citizens who sentence themselves to solitary confinement every evening as a reflection of their high level of fear of becoming a victim of violent crime should they venture outside, and the millions (perhaps *all* of us) increasingly desensitized to, more tolerant of, more adapted to, and more expectant of high levels of aggression in daily living. We most definitely do not share this perspective. Much has already been accomplished in both the laboratory and the community successfully to reduce child abuse and other crimes of violence (Steinmetz & Straus, 1974; Goldstein, Apter & Harootunian, 1984; Goldstein, Keller & Erne, 1985; Goldstein & Glick, 1987).

Towards possible solutions
The remainder of this chapter builds upon these constructive beginnings and points to a series of what we believe to be vital next steps towards the goal of aggression reduction in contemporary society. These steps, all of which require a change in our customary thinking about what

constitutes an effective means for reducing aggression, will focus upon intervention *complexity*, *prescriptiveness*, and *situationality*, and are based on the belief that aggression and its prosocial alternatives are primarily *learned* behaviors.

Complexity

There is a clear tendency, emerging especially perhaps from the 'can do' ethos traditionally characteristic of the United States, to hope for and expect to find the 'big solution' to major social problems. Thus, we seek the one breakthrough or intervention program which will wipe out poverty (e.g. the War on Poverty of the 1960s), cancer (e.g. various miracle cures), illiteracy (e.g. a sequence of early education/stimulation interventions), hunger/malnutrition (e.g. food stamps), and drug abuse (e.g. 'Say No' campaigns) and most certainly, aggression also. The 'magic bullets' aimed at aggression over the past few decades have included psychotherapy, behavior modification, medication, early 'detection', and, in America's criminal justice system, indeterminate sentences, determinate sentences, selective incapacitation, diversion from the system altogether, and many other programs. All such magic bullets miss their targets. Breakthroughs are exceedingly rare and less and less likely as the problem's complexity increases. Simply put, we would assert that complex problems will yield only to complex solutions. Every act of aggression, we believe, has multiple causes. When Johnny throws a book at his teacher (or draws a knife), it is unproductive (for seeking possible reduction of such behaviors) to explain such behavior as caused by Johnny's 'aggressive personality', or 'economic disadvantage', or other single causes. All acts of aggression grow from an array of societal and individual causes (see Table 10.1).

In broad outline, Table 10.1 seeks to highlight the complexity of cause as a way of indicating the necessity for parallel complexity of solution. The call for complexity of solution has been heard before, from community psychologists, ecological psychologists, environmental designers and systems analysts. We ourselves have ventured into this realm earlier, in the context of school violence and means for its optimal reduction. As Table 10.2 implicitly proposes, our view is that to have even modest chance of enduring success, interventions designed to reduce aggression towards persons or property in school contexts must be oriented not only towards the aggressor himself, but also at the levels of the teacher, school administration and organization, and the larger community context. Furthermore, we held, an optimally complex intervention designed to

Table 10.1 *Multiple causes of aggressive behavior*

Cause	Examples
Physiological predisposition	Male gender, high arousal temperament
Cultural context	Societal traditions, mores which encourage/restrain aggression
Immediate interpersonal environment	Parental/peer criminality; video, movie, live aggressive models
Immediate physical environment	Temperature, noise, crowding, pollution, traffic.
Person qualities	Self-control, repertoire of alternative prosocial values and behaviors
Disinhibitors	Alcohol, drugs, successful aggressive models
Presence of means	Guns, knives, other weapons
Presence of potential victim	Spouse, child, elderly, other

reduce school violence ought to seek to do so via a variety of modes or channels. Table 10.2 details this intervention strategy.

Prescriptiveness

Having developed an appropriately complex array of intervention procedures, how might they optimally be utilized? We believe success is more likely if interventions are applied in a differential, tailored, individualized or prescriptive manner. This is a viewpoint we feel is relevant to all types of aggressive behavior, although in our own work we have operationalized it most fully in intervention efforts directed toward chronically aggressive juvenile delinquents. Simple to define in general terms, but difficult to implement effectively, prescriptive programming recognizes that different juveniles will be responsive to different methods. The central question in prescriptive programming with juvenile delinquents is *which types of youth, meeting with which types of change agents, for which types of interventions, will yield optimal outcomes?* This is a view that runs counter to the prevailing 'one-true-light' assumption which underlies most intervention efforts

Table 10.2 *A multidimensional intervention strategy for school violence*

	Multi-modal interventions				
	Psychological	Educational	Administrative	Legal	Physical
Community	Programs for disturbed children	Prosocial television programs	Adopt-a-School programs	Gun control legislation	Near school, mobile home vandalism watch
School	Use of skilled conflict negotiators	Prescriptively tailored course sequences	Reduction of class size	Legal-rights handbook	Lighting, painting, paving programs
Teacher	Aggression management training	Enhanced knowledge of student ethnic milieu	Low teacher–pupil ratio	Compensation for aggression-related expenses	Personal alarm systems
Student	Interpersonal skills training	Moral education	School transfer	Use of security personnel	Student murals, graffiti boards

directed towards juvenile offenders and holds that specific treatments are sufficiently powerful to override substantial individual differences and thus aid heterogeneous groups.

Research in all fields of psychotherapy has shown the one-true-light assumption to be erroneous, especially with regard to aggressive and delinquent adolescents. Palmer (1975) reviewed Martinson's (1974) examination of a large number of diverse intervention efforts designed to alter the deviant behavior of juvenile offenders. Martinson had concluded that 'With few and isolated exceptions, the rehabilitative efforts that have been reported so far have had no appreciable effect on recidivism', (p. 25), but Palmer pointed out that this conclusion was based on the one-true-light assumption. In each of the dozens of studies reviewed by Martinson, there were homogeneous subsamples of adolescents for whom the given treatments under study had worked. Martinson had been unresponsive to the fact that when homogeneous subsamples are combined to form a heterogeneous full sample, the various positive, negative and no-change treatment outcome effects of the subsamples cancel each other out. But when smaller, more homogeneous subsamples are examined separately, interventions do work. The task then is not to continue the futile pursuit of the one approach that works for all, but instead to discern which intervention administered by which providers work for whom.

The 'many-true-lights' prescriptive programming viewpoint has its roots in analogous endeavors with populations other than juvenile delinquents (such as the treatment of emotionally disturbed adults and children) and attempts to improve elementary and secondary education. These ample precedents, however, are not the only beginnings of concern with prescriptive programs in juvenile corrections. It is worth noting, for example, the differential effectiveness of each of the two most widely used interventions with juvenile delinquents, i.e. individual and group psychotherapy. Individual psychotherapy has been shown to be effective with highly anxious delinquent adolescents, the socially withdrawn, and those displaying at most a moderate level of psychopathic behavior. Adolescents who are more blatantly psychopathic, who manifest a low level of anxiety, and those who can be described as 'non-amenable' are appropriately viewed as poor candidates for individual psychotherapy interventions.

Many group approaches have been developed in attempts to aid delinquent adolescents. Some of the more popular have been activity group therapy, guided group interaction and positive peer culture.

Research demonstrates that such approaches are indeed useful for older, more sociable and person-oriented adolescents, for those who tend to be confrontation-accepting, for the more neurotic-conflicted, and for the acting-out neurotic. Juveniles who are younger, less sociable, or more delinquent or who are confrontation-avoiding or psychopathic are less likely to benefit from group intervention. Other investigations also report differentially positive results for such subsamples of delinquents receiving individual or group psychotherapy as 'immature-neurotic', short-term rather than long-term incarceration, 'conflicted' and those 'reacting to an adolescent growth crisis'.

Other investigators imply that their non-differentially applied approach is an appropriate blanket prescription, useful with all delinquent subtypes. Keith (1984) writes in this manner as he reviews the past and current use of psychoanalytically oriented psychotherapy with juvenile delinquents. Others assume an analogously broad, non-prescriptive stance toward group psychotherapy, family therapy and behavior modification. Although in this last instance there in fact exists very substantial evidence that the sets of interventions comprising behavior modification do have very broad effectiveness, at least for the acquisition of new behaviors, the good results of behavior modification used as a blanket prescription offer the promise of even better outcomes if differential prescriptions are employed.

The utility of the differential intervention perspective with juvenile delinquents apparently also extends to interventions that are less directly psychotherapeutic and more singularly correctional/administrative in manner. There is evidence that probation, for example, may yield better outcomes for adolescent offenders who are neurotic, who display a reasonable level of prosocial behavior or social maturity or who are, in the terminology of the Interpersonal Maturity System, 'Cultural Conformists'. Probation appears to be a substantially less useful prescriptive intervention when the youth is non-neurotic, or low in social maturity.

Diversion from the juvenile justice system is a more recent correctional/administrative intervention than probation, so less opportunity has existed for its different utilization and examination. Yet even here, there may well be prescriptive exceptions to these negative combined results. Younger delinquents appear to profit more from diversion than older adolescents. Those delinquents in diversion programs for lengthy periods change more than those involved for shorter periods. Such a result is obscured when adolescents of all types or programs of all types are aggregated into large samples for overall analyses of effects.

In discussing prescriptive programming we have so far focussed on two of the variables that combine to yield optimal prescriptions: these are the interventions and the types of youth to whom the interventions are directed. But optimal prescriptions should be tridifferential, specifying type of intervention by type of client by type of change agent. Interventions, *as received by the youths to whom they are directed*, are never identical to the procedures specified in a textbook or treatment manual. The person who administers the intervention makes a difference. The intervention must be fitted to the offender: thus internally oriented change agents have been found to be highly effective with high-maturity offenders but detrimental to low-maturity offenders, and 'relationship/ self-expression' change agents achieved their best results with 'communicative–alert, impulsive–anxious, or verbally hostile–defensive' youths, and did least well with 'dependent–anxious' ones. Change agents characterized as 'surveillance/self-control' did poorly with 'verbally hostile–defensive' or 'defiant–indifferent' delinquents, but quite well with the 'dependent–anxious'. Agee (1979) reported similar optimal pairings. In her work, delinquents and the change agents responsible for them were each divided into expressive and instrumental subtypes. The expressive group contained adolescents who were overtly vulnerable, hurting and dependent. The instrumental group contained youths who were defended against their emotions, independent and non-trusting. Expressive staff members were defined as open in expressing their feelings and working with the feelings of others. They typically value therapy and personal growth and see this as an ongoing process personally and for the youths they treat. Unlike the expressive delinquent youngsters, though, they have resolved past problems and are good role models because of their ability to establish warm, rewarding interpersonal relationships. Instrumental staff members were defined as not being as comfortable with feelings as the expressive staff members were. They are more likely to be invested in getting the job done than processing feelings and are more alert to behavioral issues. They appear self-confident, cool and somewhat distant, which impresses the instrumental delinquent.

Agee indeed reports evidence suggesting the outcome superiority of (*a*) expressive–expressive and (*b*) instrumental–instrumental youth–change agent pairings, a finding in substantial part confirmed in our own examination of optimal change agent empathy levels when working with delinquent youths.

We have in this section exemplified the initial fruits of a prescriptive intervention strategy in the realm of juvenile delinquency. We have done so to reflect fully the importance we believe this perspective must be accorded in efforts to reduce aggression.

Situationality

Psychology's basic assumption, at least until the 1950s, was that the primary determinants of human behavior lie within persons themselves, i.e. their personality or dispositional tendencies. Understand the personality better and, this belief held, the accurate prediction and control of behavior will follow. But as psychology moved into the 1960s it became increasingly clear that the cross-situational consistency of behavior, which should be evident if behavior is largely determined by dispositions the person brings to diverse situations, quite often did not occur. It was shown that '(*a*) simple trait-oriented conceptions of personality must be replaced by much more *complex* conceptualizations and (*b*) the *scope* of these conceptualizations must be expanded from exclusively person-oriented conceptions to approaches which include the domains of situations, behavior, and their interactions' (Price & Bouffard, 1974, p. 579).

This call has been heeded. Modern psychology has moved away from the exclusive use of personality indices and increasingly relies on person plus situation interactional information. This perspective has found substantial expression *vis-à-vis* many different classes of behavior including aggression.

Personality information, of course, remains of major value. We can make lower-order successful predictions about future aggressive behavior, for example, if we know something about the hostility trait level of a given athlete, student or prisoner. But how much more effective our intervention efforts when we also factor in the situational research findings that

1 Aggression in an athletic context is greater for members of home teams than visiting teams, greater later in the season than earlier, later in the game than earlier, and when the team is in the middle rather than top or bottom of its league standings.

2 Aggression in a school context is greater in some physical locations (cafeteria, stairwells, bathrooms) than others, greater at some times of the year than others (March is the worst month in the USA; the last day of school is worst in Japan), greater the larger the school, and greater when the school's governance is either autocratic or *laissez-faire*, as opposed to 'firm but fair'.

3 Aggression in a prison context is greater the larger the prison, the more external (in and out) traffic, the more internal (within) traffic, and the more racially mixed or gang-dominated.

Situations have many features potentially useful – when combined with personality information – for understanding, prediction and control. These include its location, physical arrangement, entrances and exits, illumination, temperature, noise level, the people present, its time of occurrence, the actions which take place, its norms, rules, goals, roles, tasks, themes, expectations, ambiguities, and more. These and other situation characteristics are each potential antecedents or correlates of aggressive behavior, and none of them are 'in' the potential aggressor.

Aggression as learned behavior

In addition to thinking about aggression complexly, prescriptively and situationally, serious efforts to deal with human aggression require us to focus on the role of learning in its development. We will not repeat here the theoretical and empirical case against the competing, 'instinctive' view of aggression. That marshalling of negative evidence, described above in our consideration of catharsis, convincingly shows the bankruptcy of the position that aggression basically derives from instinctive energy. What, then, is a sounder alternative view? A mass of information has been gathered in recent years definitively demonstrating that a major (perhaps *the* major) source of aggressive behavior is human learning. Bandura (e.g. 1973) was an early proponent and demonstrator of this view, and many have since followed. But learned where, and how? In most western societies, and perhaps most clearly in the United States, there appear to be three major classrooms for aggression: the home, the school and the mass media. In the American home, while 'only' 4 per cent of parents physically abuse their children (e.g. burn, fracture bones, shake to point of concussion), 85 per cent make at least occasional and sometimes frequent use of corporal punishment (e.g. spank, hit, slap). What happens when an adult hits a child? The first frequent consequence is that the child ceases the behaviors which the adult perceived as aversive. Negative reinforcement is the cessation of an aversive event and, thus reinforced, the adult is now more likely to use corporal punishment in response to his or her child's next perceived transgression. But the child also has learned that 'might makes right'. Bandura and many others have shown that aggression (and all other behaviors) are learned in essentially two ways, directly or vicariously. Direct learning (the adult aggressor in our example) follows from reinforced practice, i.e. the behavior is tried and brings the desired result. Vicarious learning (the aggressed-upon child in our example) occurs by observing others behave aggressively (or use any other behavior) and receive reward for doing so. From the event sketched above, little Johnny not only learns not to pull his sister's hair again (at least not when his father is around) but – observing the nature and success of his father's aggression toward him – he, too, quickly learns 'might makes right'. He may well not counteraggress toward his father, but it should not surprise us if he goes out to play, sees a younger (or smaller) friend with an attractive toy, and proceeds forcefully to take it from him. The apparent inter-generational transmission of abusive parental child rearing practices stands in empirical support of the aggression-as-vicariously-learned theme of the foregoing fictitious father–son vignette.

The school is a second major location in which aggression is learned. Learning occurs not only through the pro-aggression influence of adolescent peer groups but also by means directly analogous to those described above for parents and their children. The teacher may often be a skilled – if inadvertent – instructor in aggression. While abusive physical punishment by parents toward their children is illegal, such is not the case between teacher and student. For example, in 43 of the states in the USA corporal punishment in a school setting is legal. Admittedly, in most of these 43 states constraining rules require the offended teacher to desist from administering the punishment herself or himself and, instead, physically escort the purported offender to the office of the school's 'designated hitter', often a vice-principal. The latter functions under his own stipulated constraints, about size of paddle, number of strokes, intensity of strokes, and so forth. Such elaborate rules and regulations may serve to make the agents of punishment feel more comfortable, but do not in our view change at all the great likelihood that when Johnny leaves the vice-principal's office, rubbing his behind, he will have had a very clear lesson that, once again, might makes right.

One does not only learn aggressive ways of behaving in the formal educational setting of schools and the less formal contexts of homes, but also perhaps especially, via the mass media, i.e. newspapers, books, comic books, radio, movies and, in particular, television. The impact of – especially America's – mass media on the behavior of its citizens is immense. One of its effects is to increase the level of violence in contemporary America. This assertion is still disputed in some quarters, but our reading of the combined evidence bearing upon the influence of television viewing (in particular, on overt aggression) leaves little room for doubt or equivocation. The very heavy, almost unremitting diet of violence on American television is a very substantial contributor to both the acquisition of aggressive behavior and the instigation of its actual enactment.

Prime-time, evening television in the United States during 1986 contained an average of 13 acts of violence per hour. The comparable figure for 1982 was seven such acts per hour. Cartoon violence, on Saturday mornings, now contains 36 violent acts per hour. By the age of 16, the average American adolescent – who views 27 hours of television per week – will have seen 200,000 attempted murders, 33,000 of which succeeded. No wonder a substantial minority of viewers will engage in actual, copy cat violence!

The pernicious effects of television violence go further, and extend to the substantial decrease in sensitivity, concern and revulsion to violence among the general viewing audience. Higher and higher levels of violence become more and more tolerable. Arguments of the television industry to the contrary notwithstanding, we believe that much of American television is appropriately viewed as pernicious and promotive of the worst in human behavior. Findings from other countries and from cross-cultural studies also demonstrate the significant role of violent TV in the occurrence and maintenance of aggression.

Yet anything learned can be unlearned, and the very means by which aggressive, antisocial behaviors are typically learned (e.g. rewarded direct experiences or observation of rewarded others) can and have been utilized to teach prosocial alternative behaviors, such as sharing, empathy, cooperation, helping and altruism. In the final section of this chapter we will illustrate this perspective by describing a learning-based intervention utilized successfully to reduce the aggressive behavior of chronically aggressive individuals. In addition, the example which follows also gives expression to the other ways of conducting intervention planning and implementation which we have described as vital steps for reducing aggression, namely complexity, prescriptiveness and situationality.

Aggression Replacement Training

Aggression Replacement Training (ART) (Goldstein & Glick, 1987) is a recently developed and evaluated intervention approach designed for use with chronically aggressive juvenile delinquents. Its three learning-based components were selected or developed to construct a multi-method, multi-channel, optimally complex intervention. These components, designed simultaneously to decrease the likelihood of antisocial behavior and increase the likelihood of prosocial, alternative behaviors, are structured learning, anger control training and moral education.

1. *Structured learning.* In this approach small groups of chronically aggressive adolescents[2] with shared psychological skill deficiencies are treated as follows:

(a) They are shown several examples of expert use of the behaviors constituting the skills in which they are weak or lacking (i.e. *modeling*).

(b) They are given several, guided opportunities to practice and rehearse these competent interpersonal behaviors (i.e. *role playing*).

(c) They are provided with praise, re-instruction and related feedback on how well their role playing of the skill matched the expert model's portrayal of it (i.e. *performance feedback*).

(d) They are encouraged to engage in a series of activities designed to increase the chances that skills learned in the training setting will endure and be available for use when needed in the school, home, community, institution or other real-world setting (i.e. *transfer training*).

Structured learning is a systematic, psychoeducational intervention demonstrated across a great many investigations to serve reliably the purpose of teaching a 50-skill curriculum of prosocial behaviors. Stated simply, in addition to other target behaviors, it teaches youngsters behaviors they may use instead of aggression in response to provocations they may experience.

2. *Anger control training.* Developed by Feindler and her research group (Feindler, Marriott & Iwata, 1984) at Adelphi University, based in part on the earlier anger control and stress innoculation research, respectively of Novaco (1975) and Meichenbaum (1977), anger control training – in contrast to structured learning's goal of prosocial behavior facilitation – teaches the inhibition of anger, aggression and, more generally, antisocial behavior. By means of its constituent components, e.g. identification of the physiological cues of anger and its external

---

[2] Consistent with a systems, ecological expression of intervention complexity, our most recent evaluation of ART with chronically aggressive juvenile delinquents has also contained ART groups designed for parents of these youths, as well as their siblings.

triggers or instigators, self-statement disputation training, refocussing anticipation of consequences, and so forth, chronically angry and aggressive youths are taught to respond to provocation (others' and their own) less impulsively, more reflectively, and with less likelihood of acting-out behavior. In short, anger control training teaches youngsters what not to do in anger-instigating situations.

3. *Moral education.* Armed with both the ability to respond to the real-world prosocially, and the skills necessary to stifle or at least diminish impulsive anger and aggression, will the chronically acting-out youngster in fact choose to do so? To enhance the likelihood, one must enter, we believe, the realm of moral values. In a long and pioneering series of investigations, Kohlberg (1973) had demonstrated that exposure of youngsters to a series of moral dilemmas, in a discussion-group context which includes youngsters reasoning at differing levels of moral thinking, arouses an experience of cognitive conflict whose resolution will frequently advance a youngster's moral reasoning to that of the higher-level peers in the group. While such an advance in the stage of moral reasonings is a reliable finding, efforts to utilize it by itself as a means of enhancing actual overt moral behavior have yielded only mixed success, perhaps, we would speculate, because such youngsters did not have in their repertoires the skills either for acting prosocially or for successfully inhibiting the antisocial. We thus reason that Kohlbergian moral education has marked potential for providing constructive directionality toward prosocialness and away from antisocialness in youngsters armed with the fruits of both structured learning training and anger control training.

We would emphasize here our basic belief that the sources and maintainers of aggression are diverse and multi-channel. So too must be its remediation. Structured learning training is our behavioral intervention; anger control training is affective in its substance; moral education is cognitive in nature. Guided by this multi-model philosophy, we have in fact been able to demonstrate that ART yields outcomes superior to those resulting from single-channel interventions. Research evaluating ART has also begun to confirm its initial prescriptive value for reducing the aggressive behavior of juvenile delinquents.

The evaluations of the effectiveness of ART should be viewed as but a prescriptive first step. We have demonstrated ART's effectiveness across relevant criteria and, to some extent, settings, but we have not begun to identify those types of youth for whom it is an ineffective technique, which types of trainers are and are not effective in its utilization, nor have we examined type of youth by type of trainer effects or youth by trainer by training method interactions. One study sought to identify inter-personal maturity level discriminators of response and non-response to ART. Differential results of this analysis did not emerge; ART was equally effective across I-level types.

Clearly, therefore, we wish to underscore for others seeking to examine the effectiveness of ART, or other interventions aspiring to serve as effective means for reducing aggression, the substantial value at this point of conceptualizing and implementing experimental designs that move beyond the non-differential prescriptive level.

We have sought to describe the ways in which ART is an optimally complex, learning-based, moderately prescriptive intervention. We have also sought to reflect in its procedures and administration our belief in the importance of situationality. Prosocial skill enactment in each participating trainee's skill role-playing is designed, collaboratively by trainer and trainee, to be idiosyncratic and relevant to where, when and with whom he will actually need to use that given skill in his or her own real-world environments. Such diagnoses of skill application are a major component of each role play's preparation, enactment and homework application in the actual setting for which it was previously rehearsed. So too with ART's anger control training component. Each participating youth is required to bring to each session a completed sheet from his 'hassle log'. These are descriptions of recently experienced conflicts. The anger-diminishing steps which constitute the anger control process are then rehearsed – in the group and outside – with particular attention to the interpersonal and environmental characteristics of the real-world setting to which the trainee will return, and in which anger control ability will be needed. As with our operationalization of prescriptiveness, what we have just presented as our implementation of situationality is but a mere beginning. We offer it not as a model, but as a directional arrow, pointing to one final quality we feel will substantially enhance the effectiveness of interventions designed to reduce human aggression.

Conclusion

We have sought in this chapter to highlight the ways in which currently employed interventions designed to reduce human aggression, largely reflecting belief in control, catharsis or cohabitation, are inef-fective and hold little potential for successful outcomes. Our intervention planning and implementation, we have held, must become more complex, prescriptive, situational and learning based. These planning and implementation perspectives certainly do not exhaust the possible new directions which aggression intervention efforts might profitably take, but in our view they are a very major beginning. Let us begin.

References and further reading

Agee, V. L. (1979). *Treatment of the Violent Incorrigible Adolescent.* Lexington, Mass.: Lexington Books.

Bandura, A. (1973). *Aggression: a social learning analysis.* Englewood Cliffs, New Jersey: Prentice-Hall.

Bartollas, C. (1985). *Correctional Treatment: theory and practice.* Englewood Cliffs, New Jersey: Prentice-Hall.

Berkowitz, L. (1964). The effects of observing violence. *Scientific American*, **210**, 35–41

Fagan, J. A. & Hartstone, E. (1984). Strategic planning in juvenile justice – defining the toughest kids. In R. A. Mathias, P. D. Muro & R. S. Allison (eds), *Violent Juvenile Offenders*. San Francisco: National Council on Crime and Delinquency.

Feindler, E. L., Marriott, S. A. & Iwata, M. (1984). Group anger control training for junior high school delinquents. *Cognitive Therapy and Research*, **8**, 299–311.

Freud, S. (1950). Why war? In J. Strachey (ed.), *Collected Papers of Sigmund Freud*, vol. 5. London: Hogarth Press.

Goldstein, A. P. & Glick, B. (1987). *Aggression Replacement Training*. Champaign, Illinois: Research Press.

Goldstein, A. P. & Segall, M. H. (eds) (1983). *Aggression in Global Perspective*. Elmsford Cliffs, New Jersey: Pergamon Press.

Goldstein, A. P., Apter, S. J. & Harootunian, B. (1984). *School Violence*. Englewood Cliffs, New Jersey: Prentice Hall.

Goldstein, A. P., Keller, H. & Erne, D. (1985). *Changing the Abusive Parent*. Elmsford, New York: Pergamon Press.

Goldstein, J. H. & Arms, R. L. (1971). Effects of observing athletic contests on hostility. *Sociometry*, **34**, 83–90.

Hagan, J. (1982). *Deterrence Reconsidered: methodological innovations*. Beverly Hills: Sage.

Keith, C. R. (1984). Individual psychotherapy and psychoanalysis with the aggressive adolescent: a historical review. In C. R. Keith (ed.,) *The Aggressive Adolescent*, pp. 191–208. New York: Free Press.

Kohlberg, L. (ed.) (1973). *Collected Papers on Moral Development and Moral Education*. Cambridge, Mass.: Center for Moral Education, Harvard University.

Martinson, R. (1974). What works? Questions and answers about prison reform. *The Public Interest*, Spring, 22–54.

Meichenbaum, D. (1977). *Cognitive Behavior Modification: an integrative approach*. New York: Plenum Press.

Novaco, R. (1975). *Anger Control: the development and evaluation of an experimental treatment*. Lexington, Mass.: D. C. Heath.

Palmer, T. (1975). Hostility catharsis: a naturalistic quasi-experiment. *Personality and Social Psychology Bulletin*, **1**, 195–7.

Price, R. H. & Bouffard, D. L. (1974). Behavioral appropriateness and situational constraint as dimensions of social behavior. *Journal of Personality and Social Behavior*, **30**, 579–86.

Ross, R. R. & Fabiano, E. A. (1985). *Time to Think*. Johnson City, Tennessee: Institute of Social Sciences and Arts.

Russell, G. W. (1983). Crowd size and density in relation to athletic aggression and performance. *Social Behavior and Personality*, **11**, 1.

Steinmetz, S. K. & Straus, M. A. (eds) (1974). *Violence in the Family*. New York: Dodd, Mead & Co.

D COMMUNICATION AND GROUP PROCESSES

Editorial

The chapters in the following section deal with aggression as a *social* phenomenon. Aggressive acts imply interaction and communication between two or more individuals or between groups: an aggressor attacks a victim, one group fights against the other, a leader gives an order to start an offensive. In addition, interpretation of the situation may be a crucial element in an aggressive interaction: one person hits another because he interprets his gesture as a threat. Thus, for the analysis of aggression it is necessary to consider both the *interaction* processes, and the *communication* and *interpretation* patterns associated with the interaction. Especially important are the cognitive processes related to interpretation, including evaluation of the context, conscious comparison between behavioural alternatives (e.g. aggressive versus non-aggressive), and anticipation of the consequences.

Interactions take place either within a dyad, or within a group involving three or more members, or between groups. As soon as a *group* is involved, the principles described for individual aggression are insufficient to explain the causes, development, or functions of aggression. Before turning to the chapters in this section, we will briefly present some basic results concerning the interaction and communication aspects of aggression.

Many studies demonstrate that an individual's behaviour in a group differs from his behaviour when alone. In this context it is necessary to distinguish between intragroup and intergroup aggression. Intragroup aggression refers to aggressive conflicts between members of the same group, intergroup aggression denotes aggressive conflicts between two or more different groups. Some of the processes involved in the latter are described in the chapter by Rabbie, which deals with the cognitive, emotional, motivational and normative orientations that can lead to intergroup conflicts.

An important factor in intragroup aggression concerns the acquisition of leadership or high status in the group. The characteristics leading to high status depend, of course, on the nature of the group: in some cases intellectual skills or prosocial behaviour may be critical. But in a group defined in terms of violence, such as a terrorist group or a criminal gang, aggressiveness may be a necessary but not a sufficient requirement for high status. The aggression criterion is especially important in groups where the leader symbolizes the goals of the aggressive group, e.g. a charismatic gang leader. In this and similar cases the common group definition, the dominant group norm, reinforces the performance of aggression in preference to alternative behaviours. Aggression depends on the history of the group and its norms.

Group members who are already leaders but whose position is threatened tend to be more aggressive towards external goals, presumably to stabilize their own position. The same seems to be true for a whole group: a decrease in cohesion within the group increases the likelihood of aggressive acts towards other groups.

Intergroup aggression is facilitated also by group pressures: individuals act more aggressively in a group than when alone to avoid isolation and/or because they are obedient. Milgram in his famous (though somewhat criticized) studies could demonstrate how individuals, under the influence of obedience, were ready to injure others up to the level of severe physical harm and even death.

Deindividualization within a group seems also to facilitate aggression. Le Bon & Zimbardo have described the violent behaviour of individuals in a crowd. It is explained in terms of an unstructured, non-normative situation, anonymity, reduced self-responsibility, mutual imitation, behavioural contagion, etc.

The mere presence of others can itself have an effect. It increases physiological arousal which in turn reinforces and increases pre-existing behavioural tendencies, including aggressive ones. Physiological arousal in a group seems to provide at least a partial explanation for the 'courageous' rushing-forward of soldiers in the battlefield or the reaction of the masses during events like the 'Nürnberger Reichsparteitag'.

But non-emotional, non-physiological reactions are also affected by the group situation. Group decisions tend to be riskier and more polarized towards the extremes of what seems to be the dominant norm of the group. If there is already a predisposition for aggressive decisions, they will become more likely in the group. This may frequently apply to staff decisions made during wartime.

A widely discussed group phenomenon related to aggression is 'groupthink'. Its symptoms are the illusion of invulnerability, the justification of past decisions, an unquestioned belief in the group's morality, self censorship, and the presence of 'mindguards', individuals who pay special attention to the group members' behaviour and possible dropout tendencies. It is more likely when there is a high need for harmony and cohesion, when the group is relatively isolated from contrary viewpoints, and if it is under high stress. The consequences of groupthink are, among other things, aggressive suppression of opposing views and stereotyping, running down, and sometimes even dehumanizing another group, an 'enemy'. Apart from mere competition for interests, territory, etc., intergroup aggression often involves such stereotyping and dehumanization. These factors facilitate aggression against an outgroup and increase the stability of the ingroup.

One major example of the ingroup–outgroup phenomena that include aggression is racism. Because racism was often a basis for war or strong collective violence, a specific chapter on its phenomenology by Santiago Genovés is included here. At the same time, the case of racism demonstrates that a long-term belief about a supposedly 'natural' fact can be scientifically falsified and thus changed. It therefore serves as a model for the discussion of violence and war as 'natural'.

Racism usually consists of prejudicial attitudes and discriminatory, often physically aggressive behaviour towards members of a specific race, where attitudes and behaviour may be inter-related in that either may lead to the other. Violent activities can be the consequence of racialist attitudes but prejudicial attitudes also follow previously existing discriminatory structures and are developed to justify aggressive acts – so-called 'Blaming the victim'. In the case of racism, aggression and prejudice can form a vicious circle. Aggression against other races supports prejudice and prejudice can legitimize and create aggression and even war.

Apart from sometimes stabilizing and legitimizing pre-existing differences in economical well-being and status, aggression against another group or race may also enhance in-group conformity, assist group stabilization, or provide a scapegoat to be blamed for frustrations like unemployment or loss of economic power. On the personality side, Adorno has assumed that the authoritarian character is a cause of racism. However, other authors like Feger postulate rather a perceived threat against the value system of the dominating group as a major cause of racism.

Psychological mechanisms similar to those involved in racism contribute also to the 'enemy image', a stereotyped perception of the 'enemy's' motivations, intentions, behaviours and characteristics usually in a pejorative way. This image of the enemy probably has two main sources. One is the fear of strangeness, a fear that can be observed during a child's first year of life and is shared in some degree by all humans. The second is related to the fact that members of any group tend to see the members of their own group as generally similar to themselves and members of other groups as different, exaggerating the differences between ingroup and outgroup and perceiving the latter is inferior. This again may help to unify the ingroup.

The enemy image is frequently mutual. It is accompanied by mirror image perceptions; the respective stereotypes of two competing enemies are often structurally very similar: 'We want peace but our enemy doesn't; therefore we must increase armament.' This of course, can create another vicious circle. As Adams (1984) and Wahlström (1987) have recently argued, the enemy image contributes to the belief that war is inevitable because it is 'natural', and both factors prevent people from believing in the possibility of non-aggressive solutions to international conflicts.

Thus, beliefs, perceptions and interpretations about the other individual or group and their intentions is a crucial element in human aggression. In this context, communication plays a major role. Common interpretations of an aggression-inducing situation are reached in face-to-face communication; social influence is part of the above processes. While *personal* communication and influence, including persuasion and propaganda through (opinion-) leaders, has probably always been present in aggressive group conflicts, this century has faced the introduction of electronic *mass communication*. Mass media can contribute to the definition of events, situations and other groups. As part of this, stereotyping may be facilitated or inhibited by the media. They may help to create enemy images, but they can also contribute to a communication that increases mutual understanding. They can put topics like violence on the public agenda and, through labelling, may either increase or decrease the perceived relevance of armed conflicts.

Until recently, however, research concerning the relationship between aggression and the media has focussed primarily on the effects of violence portrayed in the media on aggressive behaviour and attitudes.

Most of the over 4000 studies on that topic have demonstrated that there is a consistent though not necessarily high link between mass media, especially television, and the viewers' aggressive tendencies. In her chapter, Lagerspetz discusses some difficulties in research on TV violence and some of its results.

For the link between *television violence* and the viewers' aggression one has to distinguish between short-term effects, long-term effects and reciprocal processes. Short-term effects include the acquisition of aggressive behaviour through social learning: aggressive actors serve as models for viewers, especially for children. The acquired behaviour pattern may then be generalized to other contexts. While in this case the content of the film is the major factor, formal features, as described by Hertha Sturm (1987), can also have short-term effects. Fast zooms and shots, loud music and sudden changes increase the viewer's physiological arousal which in turn may facilitate the occurrence of aggression if there are already personal or situational dispositions.

Long-term effects of media content include habituation and permissiveness towards aggression. By their mid-teens, for instance, American children have already witnessed several tens of thousands of killings on the screen. As a consequence they may regard aggression as an everyday phenomenon and frequently as an adequate and common means for solving conflicts. As non-violent alternatives are not offered equally often in many media, because they are more difficult to produce in a way that is interesting and attractive for the viewer, information on aggressive behaviour is preselected and may immediately create limited cognitive sets, i.e. perception and expectation patterns, and behavioural habits.

The most likely link between the media and aggression is a reciprocal one: news media reflect societal aggressiveness but also create the image of a violent world; viewers with aggressive predispositions prefer violent content which in turn reinforces their aggressive tendencies.

While differences between animal groups can sometimes be described as 'protocultural', the cognitive capacities of humans make possible cultural differences between groups that affect the propensities of individuals for aggression. In this context, a cross-cultural approach can be used to distinguish between pancultural and environmentally or culturally determined factors in the occurrence of aggression, as Segall shows in his chapter. If a certain behavioural pattern cannot be found in one or more cultures it cannot be regarded as a universal behavioural characteristic. The conclusions reached by Segall, a social psychologist, on the interaction between biological and cultural factors mesh well with those put forward in an earlier chapter by Herbert, a physiologist.

References and further reading

Adams, D. B. (1984). There is no instinct for war. *Psychological Journal*, 5 140–44.
Groebel, J. (1983). Federal Republic of Germany: aggression and aggression research. In A. P. Goldstein & M. H. Segall (eds), *Aggression in Global Perspective*. New York: Pergamon Press.
Janis, I. L. (1982). *Victims of Groupthink: a psychological study of foreign policy decision and fiascoes*. Boston: Houghton Mifflin.
Kornadt, H. J. (1984). Development of aggressiveness: a motivation theory perspective. In R. M. Kaplan, V. J. Konecni & R. W. Novaco (eds), *Aggression in Children and Youth*. The Hague: Martinus Nijhoff.
Mummendey, A. (ed.) (1984). *Social Psychology of Aggression*. Berlin: Springer.
Sturm, H. (1987). The recipient-oriented approach. In G. J. Robinson (ed.), *Emotional Effects of Media: the works of Hertha Sturm*, pp. 45–9. Montreal: McGill University Working Papers in Communications.
Tajfel, H. (ed.) (1978). *Differentiation between Social Groups*. London: Academic Press.
Tedeschi, J. T., Brown, R. C. & Smith, R. B. (1974). A reinterpretation of research on aggression. *Psychological Bulletin*, **81**, 540–62.
Wahlström, R. (1987). The image of enemy as a psychological antecedent of warfare. In M. Ramirez, R. A. Hinde & J. Groebel (eds), *Essays on Violence*. Seville: University of Seville Press.
Zimbardo, P. G. (1969). The human choice; individuation, reason and order versus deindividuation, impulse chaos. In W. J. Arnold & D. Levine (eds), *Nebraska Symposium on Motivation*. Lincoln: University of Nebraska Press.

11

Group processes as stimulants of aggression[1]

JACOB M. RABBIE

Group processes play a major role in the occurrence of aggression and violence. This chapter deals with different possible bases of aggressive intergroup conflicts. Special attention is drawn to competition between groups and the prevailing cognitive, emotional, motivational and normative orientations that contribute to these conflicts.

Intergroup competition

The differentiation of people into in- and outgroups either within or between nations is a potential source of conflict and competition, which may, in its extreme forms, erupt into open intergroup violence and aggression. One of the main questions has been why group members have a more positive attitude about their own ingroup and its members than about an outgroup and its members. This ingroup bias (Rabbie & Horwitz, 1969), ethnocentrism, ingroup favouritism, or ingroup–outgroup differentiation is an almost universal phenomenon. The only important exception is that members of a powerless minority group may often have a more positive attitude about the dominant majority group than about themselves: sometimes members of so-called underprivileged minority groups like Jews and women apply the prejudices of the majority about their own group to themselves, leading to self-hatred and feelings of inferiority. This ingroup–outgroup differentiation has been demonstrated in many field studies, but in these studies

[1] The research described in this chapter was made possible through grants by ZWO (The Netherlands Organization for the Advancement of Pure Research, Grants 57–07; 57–97; 560–270–012). This chapter is based on a shortened and revised version of an invited address to the Congress of European Psychologists for Peace, Helsinki 1986 (Rabbie, 1987).

one is never sure whether the ingroup bias is due to specific socialization practices or explicit indoctrination to induce a more favourable attitude about the ingroup than about the outgroup. Patriotic appeals like *'Deutschland, Deutschland über alles'* or 'Right or wrong, our country' may illustrate the point. If one wishes to exclude these possible sources of contamination which occur in national groups and to examine the minimal conditions under which ingroup bias may occur, we should study these issues in the laboratory.

In an early experiment (Rabbie & Horwitz, 1969) eight subjects, either all males or all females, were randomly categorized as members of a blue group or of a green group, which were seated at both sides of a table. The four members of the green group had a green label with their name on it, had to write with green ballpoints on green paper and were consistently addressed by the experimenter as members of the green group. The same thing happened, of course, to the members of the blue group. Under the guise of a study on first impressions, the subjects were asked to rate the members of the ingroup and the outgroup on a variety of personality attributes which were scaled along a favourable/unfavourable dimension. The subjects in each of the groups had never seen each other before and were not allowed to interact with each other. In these minimal group conditions there was a slight non-significant tendency for ingroup members to rate the own group and its members more favourably than the outgroup and its members. In various replications of this study a significant ingroup bias was found, especially on the social–emotional or relational dimensions of the personality attributes which were used.

The finding that social categorization, in and of itself, may lead to a significant ingroup bias has been found by many other researchers, including Tajfel and his associates. Tajfel & Turner (1979) believe that the relations between groups are 'essentially competitive'. If they are right, this would have serious implications for the relations between groups, including the relations between nation-states. According to their social identity theory, individuals are motivated to achieve or maintain a positive 'social identity' (i.e. that part of their self-concept which is derived from the groups to which they perceive themselves to belong). By trying to achieve superiority on some valued dimension over another relevant comparison group, they try to enhance or maintain a positive social identity. These types of social or relational competition should be distinguished from realistic or instrumental competition in which individuals or groups try to achieve a material or economic advantage over each other. Our assumption is that groups are not inherently more competitive than individuals, but that depending on the dominant psychological orientations in the group or nation and on the nature of the interdependence structure between them, they may be more or less competitive than individuals. We will return to this issue later.

With respect to the social environment, we make a distinction between the internal and external social environment. Sometimes international conflicts can be better understood by focussing on the internal or domestic conflicts of interests within the nations than on the tensions between them. For example, when President Nixon alerted his troops during the October war between Israel and Egypt in 1973, there was widespread speculation in the world Press that by this move he was trying to divert the attention of the public from his controversial role in the Watergate scandal to the international conflict with the Soviet Union in an effort to unite the nation behind him.

This hypothesis, that political leaders will engage in international conflicts in order to unite the nation behind them, is very popular but lacks empirical support. That is the reason why we tried to test this hypothesis under more controlled laboratory conditions in three experiments (Rabbie & Bekkers, 1976). In these experiments, leaders were given the choice of opting either for intergroup competition or for intergroup cooperation when they were threatened with the loss of their leadership position by a majority vote of their constituents. Consistent with our expectation, threatened leaders appeared to have a greater preference for intergroup competition than for intergroup cooperation, as compared with non-threatened leaders, especially when the factions within their group were sharply divided among each other and when they had a stronger negotiating position than another rival group. Apparently, a threatened leader is likely to chose intergroup competition when he sees some chances of bringing the intergroup negotiations to a successful conclusion. When leaders felt very weak and threatened in their leadership position, they seemed less concerned about the strength of their negotiation position. In these conditions, they showed a greater preference for intergroup competition than for intergroup cooperation, even though the chances of success were very slim indeed.

These findings remind one of the invasion of the Falkland Islands by the ruling Junta of Argentina. On the evening before the invasion, there were great demonstrations in Buenos Aires protesting about the draconic economic measures of the Government. According to the views of Hastings & Jenkins (1983) 'that night the streets erupted not to the exultant cries of war fever but to anti-government demonstrations and mob violence of a ferocity not seen in Argentina since before the military coup of 1976. It made the junta's decisions final. As La Prensa had prophetically remarked a month earlier "the only thing which can save this Government is a war (1983)".' This instrumental strategy for uniting the people behind them was very successful, at least in the short run. After three days Galtieri, the leader of the junta, could announce that the British garrison had capitulated and that the new Governor of the 'Islas Malvinas' had been appointed. People were enthusiastic about his emotional speech, which was delivered from the Presidential Palace. Again in the words of Hastings & Jenkins (1983, p. 93), 'Three days earlier, his police has been shooting at civilians in the same Plaza de Mayo. Now the square was filled with men and women weeping tears of joy.'

The notion that intragroup tensions and conflicts may contribute to intergroup or international conflicts is not incompatible with the idea that intergroup conflict depends in part upon the nature of the interdependence structure between the parties. Both determinants may be important at the same time. The arms race between the superpowers provides a good example.

There have been two major explanations for the arms race. One approach stresses the importance of internal, domestic determinants and believes that weapon manufacturers and other interested parties, which have been called by Eisenhower the Military–Industrial Complex or MIC, determine the scale of the national weapons arsenal. The other explanation emphasizes the nature of the interdependence structure between the parties and the importance of the threat that the other party may exploit one's own weakness if one is willing to disarm. We will describe that latter point of view in more detail.

Fig. 11.1. A Prisoner's Dilemma Game: a representation of the arms race between the superpowers. C = cooperation; D = defection.

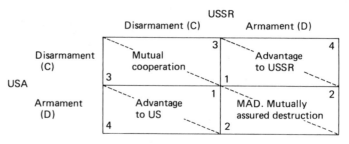

A number of game theorists have represented the presence of the arms race between the superpowers in terms of the interdependence structure of the Prisoner's Dilemma Game (PDG) (Hamburger, 1979). The matrix in Fig. 11.1 depicts the interdependence structure between the USA and the Soviet Union as they are faced with the choice between armament or disarmament. As one can see, it would be to each nation's individualistic, short-term interest to increase its level of armament while the other disarmed. In this situation, the armed country can impose its will on the other by threatening to make use of its superior military power to settle the conflicts of interests to its own advantage. In the matrix, this stage of affairs is symbolized in the lower left-hand cell by the 4 to 1 advantage of the USA and the upper right-hand cell by the 4 to 1 advantage of the USSR. If both parties decide to strive for the maximization of their own individualistic interests, without any concern about the welfare of the other, they will decide to retain their armaments and will end up in the DD-cell of the matrix. In this situation, they are worse off than if they had decided to cooperate with each other, trying to reduce the level of armaments. In the interdependence structure of the PDG, there is a clash between the individualistic and collective rationality. If each party tries to maximize its own short-term individualistic interests, both are worse off collectively than if they take a long-range perspective and decide to cooperate with each other. Game theorists are pessimistic about the outcomes of the arms race if they look at the options open to both parties. If the USSR would disarm, it is to the advantage of the USA to retain its arms. If the USSR increases its level of armaments, then the USA will be best off if it follows suit. Irrespective of what the USSR does, the USA will therefore opt for the alternative of increasing the level of armaments. The same kind of reasoning applies, of course, to decisions by the USSR. In this decision-making situation, the mutual continuation of armaments would be the inevitable result, in the absence of a world government empowered to enforce a disarmament agreement by the two nations. This analysis highlights the importance of trust, i.e. the expectation that the other is willing to cooperate. If the other cannot be trusted to cooperate, any unilateral move to disarm invites the danger that the other party may take advantage of one's own weakened position.

Bases of intergroup conflicts

A variety of cognitive, emotional, motivational and normative orientations form the bases of intergroup conflicts, and contribute to the escalation or de-escalation of these conflicts. These are considered in the following sections.

Cognitive orientations

Cognitive orientation is defined by Deutsch (1982) as a 'structure of expectations that helps the individual cognitively in the situation confronting him' (p. 24). Cognitive orientations imply 'primary' and 'secondary' appraisal processes. Primary appraisal processes refer to the expectations about the risks and opportunities that are provided by the external environment, while secondary appraisal processes are involved in the expectations of the individual or group about their capability of coping with the uncertain contingencies in the environment. Thus, the cognitive primary appraisal processes refer to the costs and benefits analyses, including beliefs about the likelihood that a given behaviour will lead to certain outcomes. Secondary appraisal processes denote the coping capability or power of the person or group. Power can be defined as the potential to influence the environment, influence as the actual exercise of power and control that is effective, i.e. effects that diminish the distance between a present state and a desired end state or goal. In attempts to predict and control the external world, people will use schemata, scripts, causal attributions and social categorization to reduce the uncertainty in their environment.

In our research, we have been particularly interested in the effects of social categorization on the way we construct and organize our social world. Sometimes the accuracy of our social categorizations can be a matter of life or death. In times of war, the social categorization of a soldier as 'one of us' or 'one of them' might be of crucial importance! Earlier it was shown that the categorization of people into blue and green groups induced an ingroup bias or ingroup–outgroup differentiation. This finding raises the question of whether the ingroup–outgroup differentiation is due to a more positive evaluation of the ingroup, to a devaluation of the outgroup, or to both tendencies at the same time.

Since Sumner coined the word 'ethnocentrism' around the turn of the century, a common assumption among social scientists has been that ingroup cohesion and outgroup hostility are invariably related to each other. In his own words: 'the relation of comradeship and peace in the we-group and that of hostility towards other groups are correlative to each other' (Sumner, 1906, p. 12). This position implies a negative correlation between the judgements about the ingroup and those about the outgroup. In our research (Rabbie, 1982), the opposite was found.

The correlations between the ingroup and outgroup rating were not negative as Sumner would have expected but positive, although not always significant. These positive correlations were somewhat higher for groups with a cooperative than for groups with a competitive orientation to each other.

Similar results have also been found in other research. Brewer, in a review of ingroup and outgroup differentiation in minimal groups, comes to a very similar conclusion. She writes 'ingroup bias rests on the perception that one's own group is better although the outgroup is not necessarily deprecated' (Brewer, 1979, p. 322). Her position is congruent with our view, that ingroup cohesion and outgroup hostility may vary independently from each other. The more positive attitude about the ingroup is dependent on the degree to which one identifies with the ingroup. According to our unit formation hypothesis, ingroup identification will be strengthened by unit forming factors, which include a common fate, physical proximity and a shared territory, positive interdependence between the members of the social category, and cooperative intragroup interaction and shared success experiences, even though one's own contribution to the success of the group may be minimal. Outgroup hostility occurs when the other group appears to frustrate the goal attainments of the ingroup in an illegitimate way. It can also act to stabilize an otherwise decreasing intragroup cohesion as was demonstrated in a study of terrorist groups by Groebel & Feger (1982).

Categorizing people into ingroups and outgroups leads to an exaggeration of the perceived differences between the groups and a minimization of perceived differences within each group. Generally, members of an outgroup appear to be less differentiated than members of one's own group. For example, Chinese may look all alike to westerners, while we make subtle differentiations between people in our own community. The Chinese appear to have the same problem with us. In the same way, we tend to underestimate the differences of opinions within the Communist bloc, while we are highly sensitive about the shades of opinion within the Western Alliance. We also tend to have less complex cognitions about outgroup members than about ingroup members. It has been noted that once a person is categorized, the person becomes just another example of the relevant schema and so is seen as much like everyone else who fits that schema. If the person fits an outgroup schema, the fit is seen as particularly tight, since outgroup schemata are less variable and less complex than ingroup schemata (Fiske & Taylor, 1984). Therefore, members of different groups often have stereotypic views about one

ich are sometimes very similar. An American social psycho-
Bronfenbrenner, visited the Soviet Union during the summer
w months after the U–2 spy incident. In talking with ordinary
: realized suddenly that the Russians' distorted picture about
ins as aggressive, untrustworthy, irresponsible and ruled by
governments that exploited and deluded the people were surprisingly
similar to the view of the Americans about the Russians: a mirror-image
as he called it. He pointed out that the danger of such a distorted
mirror-image is that 'it impels each nation to act in a manner which
enhances the fear of the other to the point that even deliberate efforts to
reverse the process are reinterpreted as evidence of confirmation'
(Bronfenbrenner, 1986, p. 19).

The existence of stereotypic mirror-images of each other has also been
found in the famous summer camp experiments of Sherif (1966) and by
the intergroup research of Blake & Mouton with managers of industrial
firms. Apparently, when well organized groups are in conflict with each
other and feel frustrated in the achievement of their goals, stereotypic
mirror-images of each other may emerge that are not conducive to
resolving the intergroup conflict in a peaceful way. The mirror-image
hypothesis fits the conflict-spiral model in international relations in which
the military preparations of one nation to defend itself against the other
are interpreted as offensive actions which require defensive measures
which in turn will be experienced as a threat etc. Deutsch (1986) has
written eloquently about the malignant-spiral process of hostile interaction between the US and the USSR and the way this vicious spiral
might be reversed. Most of his work is based on experimental research
and case studies of international relations.

Emotional orientations

Any conflict about important issues in which one party appears
to gain an unfair advantage at the cost of the other, especially if it is
obtained by illegitimate means, arouses strong emotional reactions. In
our work on individual and group aggression, we have been interested in
the question to what extent the emotional arousal which can occur in
groups may interfere with the primary and secondary appraisal processes
which are required for the cool and rational decisions necessary to
maximize or optimize the outcomes of the group or nation. Most
decisions about war and peace are made by small groups rather than by
single individuals, often in the context of a huge bureaucracy.

Influential political scientists have been rather sceptical about the value of psychological research for the explanation of international relations. These theorists advocate the 'rational actor model' in which foreign policy decisions are viewed as a product of cool, rational information processing with the aim of determining the optimal policy. Tetlock (1983) has pointed out that the 'rational actor model' is not very realistic. In making decisions about foreign policy, as in many other situations, it is very difficult to come to the right conclusions. Policy makers have to deal with incomplete, ambiguous information about the intentions and capabilities of the other states they have to deal with. The different behavioural alternatives available cannot be contemplated all at the same time. The possible consequences of each choice are shrouded in uncertainty. The policy makers in these group decisions have to make trade-offs between different values and alternatives which can hardly be compared with each other. One has to weigh economic losses and gains against international prestige, credibility and face-saving, the influence of international pressure groups, protesting allies, and internal domestic tensions. Moreover, especially in times when the outbreak of a war may be imminent, the decision makers have to function in highly stressful crisis environments in which they need to analyse large volumes of ambiguous and inconsistent evidence under severe time pressure, always with the knowledge that miscalculations may have serious consequences for their own careers and vital national interests. In these kinds of circumstances, it is not surprising that 'groupthink' may occur, which is characterized by Janis (1982) 'as the deterioration of mental efficiency, reality testing and moral judgment that results from ingroup processes' (p. 76). His analysis is based on fiascoes such as the group decisions about the invasion of the Bay of Pigs in Cuba, the decision of Truman to follow the catastrophic advice of General McArthur to invade China during the Korean war, the failure to detect the advance warnings about the impending attack of the Japanese on Pearl Harbour, and the decisions of Presidents Johnson and Nixon to escalate the war in Vietnam.

The problem with these historical examples is that accurate information is lacking about the details of the decision-making process in the group. The information is often based on the accounts of participants in the group decisions who often have strong needs to put themselves in a better light. Most such examples give rise to *post hoc*, after-the-fact explanations. Negative outcomes are attributed to groupthink, while positive outcomes are ascribed to the superior decision making of the group. In such case studies, the symptoms and antecedents of groupthink cannot be distinguished from each other. For example, it is uncertain whether the so-called antecedents of groupthink, such as ingroup cohesiveness, directive leadership and conformity pressures, should not rather be considered as consequences of severe intergroup conflicts, as the work of Sherif & Sherif and Blake & Mouton indicate.

In a recent experiment, we tried to find out whether greater emotional arousal in groups or dyads as compared with single individuals would make it more difficult for groups to suppress their angry aggression in a cool and instrumental effort to avoid retaliation from a more powerful party. By means of a Power Allocation Game, subjects, either male individuals or groups, gave the other programmed party the power to allocate money between the parties in any way they wanted. In the first two trials, the programmed party distributed the money in a fair and equal way. At the third and critical trial, however, the programmed party abused his power by giving much more to himself than to the subjects. The subjects, individuals or groups, could react to the norm violation of the other by administering painful white noise to the other over his earphones. In general, dyads reacted with more angry aggression to the norm violation of the other party than individuals did. Moreover, in an instrumental condition, in which the suppression of aggression was instrumental in not antagonizing the other in the hope that they would receive valuable rewards, less aggression occurred than in the reactive condition in which the inhibition of aggression was not seen as an instrumental response to avoid retaliation from the other.

It was expected that groups would experience a higher level of emotional arousal, making them less capable of suppressing their aggression for instrumental reasons than individuals. Consistent with these expectations, the difference in aggression between individuals and groups was much greater in the instrumental condition than in the reactive condition. However, groups may be not only less capable but also less willing than individuals to restrain their aggression for instrumental reasons. According to their questionnaire responses, groups, as compared with individuals, had stronger needs to assert themselves, were less willing to placate the other party, were more inclined to use coercive power in an effort to influence the other and were more convinced than individuals of the inherent morality of their behaviour. It should be stressed once more that, in our view, groups are not inherently more competitive and aggressive than individuals as is sometimes asserted, but that this depends upon the nature of the prevailing psychological orientations in the group.

Motivational orientations

In the context of experimental games such as the Prisoner's Dilemma Game (PDG), several motivational orientations have been distinguished, such as maximizing one's own gain (max. own), cooperation or maximizing joint gain, competition or maximizing relative gain (max. rel.), equality (minimizing differences), minimizing a maximal possible loss (or defensive motives), authorism (max. other), etc.

A distinction can be made between instrumental and relational cooperation and competition. Instrumental cooperation and competition refer to strategic responses made in an effort to maximize one's own outcomes. Instrumental cooperation between parties may occur, not because they like each other so much, but because, given a particular interdependence structure, both parties are better off cooperating than competing with each other. Our discussion about the mutual disarmament of the two superpowers is a good example of the instrumental, strategic way of thinking. The instrumental or strategic modes of thoughts and action should be distinguished from a more social or relational way of behaving. Social or relational cooperation is aimed at achieving a cooperative unit-relation with the other as a goal in itself rather than as a means or instrument to attain a goal external to the relationship. Social or relational competition is aimed at a differentiation of oneself from the other, e.g. to achieve a positive social or group identity (Tajfel & Turner, 1979). It can be argued that in the anonymous, impersonal environment of the PDG, conventional attitudes and most social motives play a minor role in behaviour. In our research, we have found, however, that in the PDG parties are motivated to cooperate with each other in an effort to establish a friendly mutually satisfying relationship or to try to beat or to maximize the difference between themselves and the other in an effort to gain more prestige and status over the other party. Social cooperation is facilitated when one sees the other as similar to him or herself and when face-to-face contact between the parties can be expected. Instrumental and relational cooperation tend to reinforce each other and this underlines the significance of good interpersonal relationships between representatives of conflicting groups. For that reason, it is very important that leaders like Reagan and Gorbachov should develop good interpersonal relationships with each other.

To find out whether group members would follow a more instrumental strategy than individuals, in the absence of intragroup interaction, socially isolated individuals were categorized as members of a green group and urged to present their group's interests as well as they could. By playing against a programmed opponent who followed a cooperative strategy, it was found that group representatives gave more instrumental reasons for their cooperative behaviour than individuals did, although they did not differ in their actual choice behaviour from individuals. In other very similar research, it has been shown that group representatives faced with a real-life responsive opponent were significantly more cooperative than individuals who represented only their own individual interests. These results suggest that intragroup interaction is not a necessary condition for group representatives to show more instrumental cooperation than individuals, provided that the other responds positively to a cooperative strategy. These findings also indicate that group representatives are not always more competitive in their negotiations than individuals are, as is often claimed in the literature. Rather, this will depend on the nature of the physical and social environment, the interdependence structure between the parties, and the prevailing psychological orientations and meaning systems of the groups and individuals. The only thing we can say is that group representatives, especially when they can be held accountable for the results of their negotiation behaviour, seem more strongly motivated than individuals to maximize the interests of their group and will either compete or cooperate depending on their beliefs as to which strategy is more instrumental in achieving that goal.

Normative orientations

Norms are rules and regulations that govern the behaviour of group members. Norms have a moral, prescriptive quality: they refer to obligations, rights and entitlements people may have in particular relations. When these norms are violated, people are likely to react with anger and aggression in an effort to rectify the injustice. It has been proposed that men's eagerness to see transgressions adequately punished derives mainly from the need to create stable, predictable environmental conditions in order to ensure that every member of a group or community can operate effectively. Norms may also have a descriptive informational function: they provide group members with evidence about social reality. Social norms differ along various dimensions from each other: whether they specify the latitudes of acceptance and latitudes of rejection of attitudes and behaviour, whether they are central or peripheral to the vital concerns of the group and whether they are explicitly verbalized or implicitly understood, whether they are sanctioned by unspecified expectations or by well defined rewards and punishments, whether they provide general direction or specific guidance, imply a broad consensus or are accepted only by a few members in the group. Explicitly verbalized norms, which are sanctioned by well defined rewards and punishments which provide specific guidance, are central to the vital concerns of the group and are based on a broad consensus in the community, become part and parcel of the external and social environment. Through socialization practices, norms may become internalized as normative orientations which guide one's attitudes and behaviour in specific situations.

In our research, we have used social norms as independent as well as dependent variables. The violation or non-violation of norms has been employed as a means to arouse angry aggression in individuals and groups composed of men and women. For that reason we have used transgressions of contractual norms, e.g. the breaking of promises, the violation of justice norms based on equity, equality or need, and the intentional abuse of power to put the other at a disadvantage. In other laboratory and field experiments, we have tried to examine the kinds of norms which may occur in specific group or crowd situations. The specific norms that emerge depend in part on the internalized normative orientations of the people involved. In various studies of group polarization by Minnix (1982), subjects were asked to indicate before and after a group discussion what the USA should do to protect its interests in several scenarios of international crisis. According to one scenario, the Naval Forces of the Soviet Union were ready to close off the Straits of Hormuz in the Persian Gulf, threatening to cut off the vital oil supplies of the USA and other members of the Western Alliance. The subjects had to play the role of members of the National Security Council of the USA. They had to advise the President what kind of measures should be taken to counter the threat. Before and after the group discussions, they had to fill out a 10-point scale which varied from rather cautious recommendations to extremely risky ones. Cautious recommendations included the call for direct negotiations with the Soviet Union, an appeal to the Security Council of the United Nations, the withdrawal of the Ambassador to the Soviet Union; the more risky measures included the institution of a blockage of naval forces, the use of conventional forces, bombing military and civilian targets, and finally the use of nuclear arms. There were three groups of subjects: officers of the American Army, ROTC cadets from the Army and Navy, and college students of the University of Cincinnati. In all three groups, strong polarization effects did occur, i.e. after the group decisions the members were more extreme in their recommendations than they had been prior to the group discussion.

These findings are consistent with our enhancement hypothesis which predicts that intragroup interaction enhances the psychological orientations within the group in the direction which already existed in the individual members prior to the intragroup interaction. The group discussions of the officers and to a lesser extent the discussions of the ROTC cadets enhanced the readiness to recommend the more risky solutions to the international conflict which involved the use of the armed forces. The members of the student groups on the other hand became more cautious in their advice to the President after the group discussion than they had been before. According to Minnix, the polarization effects are due to the dominant group norms in these different groups. International crises involve a great deal of uncertainty about the environment which can be reduced by conforming to the prevailing norms in the group. Although the laboratory situations are quite different from the actual decision-making processes by the top decision makers at the highest level of government, it is not too far-fetched to assume that the composition of the policy-making group may be important in the readiness of the group to take risks (Rabbie & Lodewijkx, 1985).

Conclusion

This paper goes only a little way in describing how research in experimental social psychology can contribute to an understanding of the problems of war. Much more could also be written about its possible contributions to the resolution of international conflict and to peace. Perhaps, in collaboration with other scientists, we can come to a deeper understanding of these issues and thus influence the policy makers on whom our lives depend. Although the contribution of experimental social psychology to gaining insights into the momentous issue of war and peace can be only a relatively small one, we hope that, in cooperation with other disciplines, it will contribute to the search for ways to prevent a nuclear war.

References and further reading

Austin, W. G. & Worchel, S. (eds) (1979). *The Social Psychology of Intergroup Relations*. Monterey, California: Brooks/Cole.

Ajzen, I. & Fishbein, M. (1980). *Understanding Attitudes and Predicting Social Behavior*. Englewood Cliffs, New Jersey: Prentice Hall.

Blake, R. R. & Mouton, J. S. (1961) Reactions to intergroup competition under win-lose conditions. *Management Science*, 7, 420–5.

Brewer, M. B. (1979). Ingroup bias in the minimal intergroup situation: a cognitive–motivational analysis. *Psychological Bulletin*, **186**, 307–24.

Bronfenbrenner, U. (1986). The mirror-image in Soviet–American relations. In R. K. White (ed.), *Psychology and the Prevention of Nuclear War*. New York: New York University Press.

Deutsch, M. (1973). *The Resolution of Conflict. Constructive and Destructive Processes*. New Haven: Yale University Press.

Deutsch, M. (1982). Interdependence and psychological orientation. In V. J. Derlega & J. Grzelak (eds), *Cooperation and Helping Behaviour*. New York: Academic Press.

Deutsch, M. (1986). The Malignant (spiral) process. In R. K. White (ed.), *Psychology and the Prevention of Nuclear War*. New York: New York University Press.

Fiske, S. T. & Taylor, S. E. (1984). *Social Cognition*. New York: Random House.

Groebel, J. & Feger, H. (1982). Analyse von Strukturen terroristischer Gruppierungen. In *Analysen Zum Terrorismus Band 3: Gruppenprozesse*. Opladen: Westdeutscher Verlag.

Hamburger, H. (1979). *Games as Models of Social Phenomena*. San Francisco: W. H. Freeman.

Hastings, M. & Jenkins, J. (1983). *The Battle for the Falklands*. London: Pan Books.

Janis, I. L. (1982). *Victims of Groupthink: a psychological study of foreign-policy decisions and fiascoes*. Boston: Houghton Mifflin.

Minnix, D. A. (1982). *Small Groups and Foreign Policy Decision Making*. Washington: University Press of America.

Morgenthau, H. J. (1973). *Politics among Nations: the struggle for power and peace*. New York: Knopf.

Pruitt, D. G. & Rubin, J. C. (1986). *Social Conflict: escalation, stalemate and settlement*. New York: Random House.

Rabbie, J. M. (1982). The effects of intergroup competition and cooperation on intra-group and intergroup relationships. In V. J. Derlega & J. Grzelak (eds), *Cooperation and Helping Behaviour. Theories and Research*. New York: Academic Press.

Rabbie, J. M. (ed.) (1987). Armed conflict: a behavioral interaction model. In *European Psychologists for Peace*. Proceedings of the Congress of European Psychologists for Peace, Helsinki 1986.

Rabbie, J. M. & Bekkers, F. (1976). Threatened leadership and intergroup competition. *European Journal of Social Psychology*, **31**, 269–83.

Rabbie, J. M. & Horwitz, M. (1969). The arousal of ingroup and outgroup bias by a chance to win or lose. *Journal of Personality and Social Psychology*, **69**, 223–8.

Rabbie, J. M. & Horwitz, M. (1982). Conflicts and aggression among individuals and groups. In H. Hirsch, H. Brandstätter & H. Kelley (eds), Proceedings of the XXIInd International Congress of Psychology, Leipzig, DDR, No. 8, *Social Psychology*. Amsterdam: Noord-Holland Publishing Company.

Rabbie, J. M. & Lodewijkx, H. (1985). 'The enhancement of competition and aggression in individuals and groups'. Paper presented to the Symposium Contribution of Psychology to Understanding and Promoting World Peace at the XXIII International Congress of Psychology, Acapulco, Mexico, 2–7 September 1984.

Sherif, M. (1966). *Group Conflict and Cooperation: their social psychology*. London: Routledge & Kegan Paul.

Sumner, W. G. (1906). *Folkways*. New York: Ginn & Co.

Tajfel, H., Billig, M. G., Bundy, H. P., & Flament, C. I. (1971). Social categorization and intergroup behavior. *European Journal of Social Psychology*, **1**, 149–78.

Tajfel, H., & Turner, J. (1979). An integrative theory of intergroup conflict. In W. G. Austin & S. Worchel (eds), *The Social Psychology of Intergroup Relations*. Monterey, California: Brooks/Cole.

Tetlock, P. E. (1983). Psychological research on foreign policy: a methodological overview. In L. Wheeler (ed.), *Review of Personality and Social Psychology*, **4**. Beverley Hills, California: Sage.

12

The myth of racism[1]

SANTIAGO GENOVÉS

An instinct is neither true nor false. One cannot argue with it, although much can be discussed about it. By contrast, a prejudice may be a false idea, but it is nevertheless an idea. Accordingly, we argue about it. One of the most deeply rooted prejudices that hinders peaceful relations among men is racism. This prejudice goes back to ancient times. Aristotle, based on climatic considerations, believed that the Nordic people, precisely those who would nowadays feel proud of their 'Arian blood', had been born to be slaves. Vitruvius was of the same opinion. Cicero, who stated that 'men differ because of their knowledge but not because of their ability to learn', let himself be overcome by this prejudice when asserting that the Celtic people were 'stupid and incapable of becoming educated'. Pharoah Sesostris III (1887–1849 BC) raised a stela in Egypt's southern frontier, wherein it was inscribed that no black man was allowed to enter Egypt.

The Renaissance, with its explorations and contacts with other peoples, brought along a recurrence of this prejudice. Very often strong disputes about the equality or inequality of men arose. In the sixteenth century, John Major, a Dominican Friar born in 1510, stated that it is far better not to liberate those men, who by nature were born and intended to obey, from a servile way of life, and Juan Ginés de Sepúlveda, the famous Spanish priest born in 1550, commented on the 'inferiority and innate perversity of the American Indians', asserting that they were not 'rational beings' and that the Indians were 'as different from the Spaniards as ... monkeys from men'. On the other side, there were characters such as Fray Bartolomé de las Casas, a Spanish Dominican missionary who, arriving in Mexico in 1502, defended the natives so

[1] An extended version of this paper is available from the author.

much that he has gone down in history as 'the Apostle of the Indians'. He fought strongly against this theory, maintaining that all the peoples of the world were made up of the same stuff and pleading for the abolition of slavery among Indians and black men alike, 'since reason is the same with them as with the Indians'. Of a similar kind were men like Montaigne, the French thinker, moralist and traveller, born in 1533, who observed that everyone calls barbarian those who do not accord with his own habits, and Thomas Jefferson, who often praised the North American Indian. Such prejudice is, in any case, quite often reciprocal. 'In the Great Antilles, some years after the discovery of America, while the Spaniards were sending special commissions to investigate whether the Indians had a soul or not, the latter were very busy submerging white prisoners in water in order to find out, through exhaustive observation, whether their corpses were subject to putrefaction or not' (Lévi-Strauss, 1960).

Later on, although Voltaire, Rousseau, Buffon and many others defended equality among men, other intellectuals, such as Hume, held that black men were inferior to those of white race. In Europe this prejudice prevailed in spite of the Christian tradition, so clearly antiracist: 'And He made all mankind be born of one blood, in order for them to people the earth' (Acts of the Apostles, xvii, 26). 'There is no longer Jew or Greek, nor is there slave and free man; there is no longer man and woman, because all of you are one in Christ' (Galatians, iii, 28). Notwithstanding the above, in 1772 Reverend Thomas Thompson published a leaflet entitled *How the trade of black slaves in the coast of Africa respects the principles of humanity and the laws of the revealed religion*, and Reverend Josiah Priest, in 1852, published another leaflet entitled *Bible defense of slavery*. As recently as 1900, C. Carrol was responsible for the appearance of a work entitled *The black man as beast or in God's image*, in which he stated firmly that every scientific work proved that the constitution of black men was like that of apes. On the other hand, Pope Pious XI condemned racism and an encyclical from Pious XIII, issued in 1938, considered it an apostasy contrary, in its spirit and doctrine, to Christian faith. But the power and depth of racial prejudice is indicated by the fact that Abraham Lincoln himself, that greatest of champions for the abolition of slavery, was recorded as saying: 'There is a physical difference between the white race and the black one, which I believe will always prevent both races from living together in terms of social and political equality' (speech in Charleston, 8 September 1858).

In the nineteenth century, just a few years before the appearance of *The Origin of Species*, Gobineau, a pseudo-scientist of the time, published his *Essay on the Inequality of Human Races*, on which Hitler based his 'Aryan' propaganda. We are all well aware of the consequences of this theory upon our times. But with the appearance of Darwinism, racism – or at least white racism – took a new course: many 'white' peoples were quite enthusiastic about Darwinism because, proclaiming the survival of the fittest, it confirmed their policy of expansion and aggression at the expense of the 'inferior' peoples.

This idea of selection and superiority, based more or less on Darwin's theory, may even be applied inside the same ethnic group: Erich Suchsland held the thesis that individuals who have not been successful in life necessarily belong to those elements of the population considered of 'inferior race', while rich individuals belong to a 'superior race' (Biasutti, 1953). 'Thus the bombardment of poor suburbs', Comas (1960) commented ironically, 'would result in a selection and improvement of the race'. Alexis Carrel, author of *The Unknown about Man*, follows an analogous reasoning in stating that proletarians and idle people are 'inferior' due to their natural inheritance and that, because of their physical constitution, they do not have the strength to fight: they have fallen so low that any kind of struggle has become useless.

It is, of course, unnecessary to comment on Hitler's racism, the terrible images of which are still in our minds. But we may mention some bizarre distortions produced by some of its fanatical adherents. Waltmann stated that Jesus had Aryan blood; in any case, he evidently was not the son of a Jew as Joseph was, since Jesus had no father(!) (Bouthoul, 1951). This same author distorted the names (and we assume their physical characteristics as well) of several geniuses produced by humanity, in order to show that they were really of Teutonic blood: Giotto, Jothe; Alighieri, Aigler; Vinci, Wincke; Dasso, Tasse; Buonarotti, Bohurodt; Velázquez, Valchise; Murillo, Moerl; Diderot, Tietroth, etc.

Ammon (1890) had earlier maintained that the dolicocephalics (like the Aryans) were 'socially superior' to the brachicephalians (like the Alpines) and that this was the reason why they were found in a higher proportion in cities than in the country and among the privileged social classes than among workers, a fact true only so far as Germany was concerned, but not with regard to Italy. There was another fact which he would have found still odder but which, of course, he took good care never to mention: the Aryans, in their turn, might find a race naturally destined to become their masters, for the black race is even more doli-cocephalic than the Arians. William II had already had anthropometric studies carried out on the German people in order to prove the 'pureness and superiority' of their Aryan race, studies which he never dared publish, as there turned out to be extensive areas where no single 'Aryan' individual could be found. Still another German investigator suggested that humanity could be divided into three groups: the pure-blooded Germans, who would enjoy every privilege, political and social; those of 'more or less Germanic blood', who would enjoy only limited privileges; and the non-Germans, who would be deprived of every political right and who would be sterilized in order to save civilization. It is very difficult to imagine that anyone can carry fanaticism to such an extreme. But there were some who did: Gunther (1929), a theoretician of Hitler's racism, seriously suggested that a line could be drawn separating 'Nordic' men from the rest of the world. Of an even more vaudeville flavor (as usual) was Mussolini's statement that there is an Italian race which is pure and of an 'Aryan–Nordic' type.

But things did not stop there. Pushed by political necessities, the theoreticians of Aryanism went so far as to assert that the Japanese were Teutons, descendants of the white Ainus people of Japan, though strongly mixed with the yellow race. This, however, did not prevent Rosenberg from writing that the latter possess all the moral and intellectual qualities of an Aryan people and even those of a Nordic one (Klineberg, 1965). 'The Japanese leaders offer the same biological guarantees as the German leaders.'

It is worthless to insist upon the absurdity of the Nazi racist thesis. They were themselves so mixed up and involved in so many contradictions that finally they declared that race is a mystical and intuitive concept, the evidence for which lies beyond any scientific demonstration. In view of the fact that a Teuton archetype blond like Hitler, as tall as Goebbels and as slim as Goering was inconceivable, the Nazis decided to state that 'a nordic soul could inhabit a non-nordic body' and that 'politics must go beyond science, supported by the fundamental intuitive truth of blood diversity among the peoples of the earth, drawing the logical consequence therefrom, i.e. the principle of leadership by the fittest'.

Against such reasoning, the most suitable answer is that given by Rimbaud, the brilliant young Frenchman, of a 'Nordic' type if there ever was one, who, almost 100 years ago, during the era of European colonial expansion and just when the ideas of biological superiority were in fashion, caustically exclaimed: 'I am of inferior race from all eternity.'

Racism, however, is not always so open and extreme, and it is often most difficult to convince those racists who rationalize their prejudices and believe they have logical and objective proofs to support them. Today, this kind of rationalization is almost always the result of distorted or misunderstood Darwinism. The conclusions they rapidly draw about the principle of natural selection are of two broad types, the harmful character of hybridization or genetic mixture, and the 'natural' superiority of some races over others.

The first group of prejudices is based on an initial false assumption, namely that there are pure races. To begin with, the concept of race is scientifically somewhat vague. What is biologically clear is that all men presently living belong to the same species and that within this species there are individual and group variations. Some of these group variations serve as a criterion to determine a classification of these groups into what we call races. But not all biologists agree as to precisely which characteristics should be taken into account. Sometimes they consider blood reactions, average height, proportionality of the members of the body, skin colour, texture and hair colour, or broadness of the nose; sometimes only some of these characteristics are taken into consideration, while sometimes new ones are added. The selection of criteria is thus in part arbitrary. The classification itself is also arbitrary. Authors differ over the number of races and subraces they recognize. In any case, the concept cannot be a static one, since populations are constantly in movement, and mixing is constantly occurring. Thus, race is an ever-changing biological phenomenon that can be understood only within an evolutionary context, through environmental and genetic parameters which remain in constant movement. And even if we did accept classification criteria, they would never allow us to find a pure race, no matter how far back in time we went, since throughout the evolution of *Homo sapiens* men from different groups have become mixed. 'Pure races in the sense of genetically homogeneous populations do not exist within the human species' (Genovés, 1965).

Sometimes, taking a single characteristic as the only distinctive feature of a race, it is easy to find, within certain groups, statistically significant proportions of individuals who present it. But as soon as we take several traits, the proportion of individuals who possess them all – within any human population – is statistically insignificant. For example, we all know, or believe we know, that the English have light eyes. False: only one out of every five Englishmen has this characteristic. To list only the outstanding contributions to the genetic composition of the English people, we have to mention those from Cro-Magnon, Nordic, Mediterranean and Alpine peoples and, concerning more recent groups, the Saxons, Norwegians, Danes and Normans. Furthermore, when the English settled in North America, the mixture increased. For their part, the black people in the United States are no 'purer' than the white: among their ancestors we find Congolese, Bantu, West-Europeans, Siberians, Mediterraneans, etc. And as far as the Jews are concerned, they were already quite a mixture even in Biblical times. Later, when they dispersed, so much miscegenation took place that most Jews possess racial characteristics more like those of the country they live in than those of other Jews. 'It is enough to compare a Jew from Rotterdam, solid, stout, red-faced, with his coreligionary in Salonica, thin and nervous with burning eyes and a pale complexion, in order to understand that what we usually call the Israeli-type possesses traits from numerous different peoples' (Comas, 1960). Thus, in spite of their statements on the intuitive evidence for races, the Nazis had to force the Jews to carry a Star of David in order to be able to distinguish them from the 'Aryans' who, supposedly, belonged to another species.

But even if we set aside the vagueness of the concept of race and the universality of human miscegenation, even if we admit that in some mysterious, mystical and intuitive way we could decide which are the 'real' races and what their 'purity' consists in, we would still have to demonstrate why it would be harmful for these 'pure' races to mix. This thesis was held by Mjonn (1922) and has found many followers in the United States, S. K. Humphrey, M. Grant and L. Stoddard among others. The data it is based upon, setting aside their genetic, typological and statistical inaccuracy, are nothing but interpretations based on the error of mistaking consequences for causes. It is obvious that wherever people with mixed racial characteristics are discriminated, they are likely to show undesirable social characteristics just because they are forced to survive under undesirable living and cultural conditions. Furthermore, discrimination and prejudice make racial mixture much more frequent among the lower social classes (Lundborg, 1921; Schreider, 1964).

From the standpoint of genetic science, all these ideas are gratuitously incongruent. Depending on the case, either endogamy or exogamy may be used to improve animal breeding, and the immediate effect of genetic mixture is often to prevent the manifestation of genetic defects of a recessive character inherent in either one of the parent races. This is somewhat like the example of the endogamic aristocratic families mentioned above. Breeding between members of similar origin certainly strengthens the 'best' or 'fittest' traits of the group, but also does the same for hereditary defects. In the long run (sometimes in the very long run), it may be said that, in principle, endogamy is more harmful, since the accumulation of the 'fittest' characteristics contributes to the survival of the group only if the surrounding milieu does not change, while the accumulation of genetic defects inevitably assures its eventual disappearance.

There is still the second aspect of the racial prejudice, the one which does not merely believe that genetic mixture is wrong and that it can bring along numerous evils, but is simply certain that some races are 'superior' to others. As recently as 1919, the humanitarian delegates who attended the Paris Conference wherein the Society of Nations was created rejected a declaration submitted by the Japanese delegation proclaiming the equality of all races. Even today, many white people still believe that black people are inferior. The most frequent expression of this prejudice concerns intelligence, despite the fact that it is now generally acknowledged that all instruments that purport to measure intelligence have a degree of culture specificity, that experiential factors play a crucial role in test performance, that no differences can be demonstrated in the very early years, and that within-group differences in test performance are almost invariably greater than between-group differences.

In conclusion, although politicians and propagandists have often played upon racial differences to further their own ends, there is no evidence whatsoever to justify the assumption of the superiority of one race over another. Any such assumption is just prejudice. We can clearly state that the tendentious utilization of evolutionary theory as a justification for the assumption of racial superiority, or for justifying war and violence, is totally without scientific foundation.

References and further reading

Ammon, O. (1890). *Anthropologische Untersuchungen*. Jena 1980.
Biasutti, R. (1953). *Le razze e i popoli della terra*, Vol. I *Razze, popoli e culture*. Torino Tipografico – Editrice Torinese.
Bouthoul, G. (1951). *Les guerres*. Paris: Presses Universitaires de France
Comas, J. (1960). Les mythes raciaux. In *Le Racisme devant la Science*, pp. 13–58. Paris: UNESCO-Gallimard.
Coon, C. (1962). *The Origin of Races*. New York: A. A. Khnopf.

Genovés, S. (1967). In S. Genovés (ed.), Race and racism. The Third Conference of UNESCO. *Yearbook of Physical Anthropology*, **13**, 270–80. U.N.A.M.-I.N.A.H.

Gunther, H. F. K. (1929). *Rassenkunde Europas. Mit besonderer Berücksichtigung der Rassengeschichte der Hauptvölker indogermanischer Sprache*. Munich: Y. F. Lehmanns Verlag.

Klineberg, O. (1965) *Psicologia Social*, Mexico: Fondo de Cultura Economica.

Lévi-Strauss, C. (1960). Race et histoire. In editors *Le Racisme devant la Science*, pp. 241–84. Paris: UNESCO-Gallimard.

Lundborg, H. (1921). Hybrid types of the human race. *Journal of Heredity*, **12**, 274–80.

Mjonn, Y. A. (1922). Harmonic and unharmonic crossings. *Eugenic Review*, **14**, 35–40.

Montagu, A. M. (1964). *The Concept of Race*. New York: Free Press.

Schreider, E. (1964). Investigations in social stratification of biological characters. In S. Genovés (ed.), *Yearbook of Physical Anthropology*, **12**, 184–209, Mexico. A.N.A.M.-I.N.A.H.

13

Media and the social environment

KIRSTI LAGERSPETZ

Compared with the previous history of humankind, western civilization nowadays is less barbarous. The level of education is higher, superstition has diminished, the powers of nature are more under the control of humans, and people on average have more chance to determine their own lives. In such a society, one would expect that the occurrence of violence would have decreased. This, however, is not necessarily the case. The level of violence has even increased in some civilized countries (see Segall, this volume). Many regard the mass media as one of the major determining factors for this. The media maintain the prestige of violence as a means for solving conflicts and correcting injustice, and teach the practice of violence through filmed examples. Most people see more violence on television, film and video than they see in their real lives.

It is sometimes argued that human beings have always been violent and aggressive, and that cultural influences cannot change that. In all societies and times – some would say – there will be those who are aggressive and those who are not. But as anthropologists have shown, the level of and the tolerance for violence, as well as the norms regulating violence, have been different in different times and are different in different cultures. For instance, during the Middle Ages what we now would call cruelty was more accepted in social life than it is now. Both the general level and the forms of violence vary in accordance with what the people living in a society value and are accustomed to.

Even some of those who admit that social conditions, as contrasted to hereditary factors, influence the level of violence and aggression never-theless maintain that mass media have nothing to do with it, but blame rather other factors such as unemployment, frustrations, child rearing

practices, early personality development, influences from the peer group
or the subculture, idealization of power, maintenance of a 'macho' ideal,
or competitiveness and lack of cooperation in the society. Which type of
influence is considered more potent depends on the sphere of experience
or the area of speciality of the person expressing the claim. It is beyond
doubt that all the above-mentioned factors are important for the development of aggression and the pursuit of violence. Aggression, like most
human behaviour, is influenced by a multiplicity of social factors.
Therefore, acknowledging the existence of one factor does not preclude
the influence of others.

Furthermore, the factors determining aggression function to a great
extent together or through each other. For instance, the portrayal of
violence on film influences the children who are looking at it not only
directly but also indirectly through their parents, peers or other people.
Mass media programmes which the person has not experienced can still
affect him/her through other people who have been acquainted with the
programmes or been told about them or their message. If, for instance,
human relationships are pictured as aggressive and violent, the viewer
may tend to see the world in the same way.

Or consider, for instance, the case of unemployment, which has been
suggested as one cause of the increase of violence. What could the
psychological mechanisms behind this kind of causation be? It is obvious
that being without work and thus without function and esteem in the
society implies many kinds of personal frustrations. Frustrations, in turn,
are powerful instigators of aggressive reactions. The presumed causal
chain here is clear enough. But in reasonably democratic societies, why
does the aggression take the form of individual violence instead of being
used to bring about a change in the situation, using different channels of
influence? There is, of course, a multitude of reasons for this state of
affairs. One is the high value attributed to individual power and
individual violence, the idealization of a man who paves his own way and
gets his rights through violence. The value is underscored repeatedly
each day in the fictional presentations shown on TV, video and film.

One argument against recognition of the effect of mass media on
violence is to claim that mass media are not an independent element, but
just reflect the society as it is, copying its prevailing values and attitudes.
It may usually be true that the mass media do not generally introduce
new views even when generous allowance is made for the creativity of the
artists who design the programmes. They frequently transmit and spread
values and attitudes that exist already, thereby serving as teachers to

millions of young people. However, in this way mass media can be agents of socialization that uphold and strengthen the attitudes and values of the society, whatever they happen to be, including aggressive ones. One consequence might be that the society becomes less resistant to change, when all its members are effectively socialized to support the same attitudes and to see human problems in a similar way. Knowledge of the social context is important for any serious study of the causes of violence. The complex causal network operating in any particular social setting must be recognized.

In this chapter, I attempt to discuss some of the research on the effects of mass media on violence and aggression, and to evaluate some of the evidence for a causal relationship. In addition, I shall discuss the psychological mechanisms through which the viewing of violence can be expected to affect the aggressiveness of the viewer. Violent films also cause anxiety and fear in the viewers. In this chapter, however, I will not take up the research which deals with anxiety and fear but will concentrate on the effects of the mass media on aggressiveness and violence.

Some authors, such as Freedman (1984), do not agree with the conclusion that the media have a strong effect on aggression primarily on the grounds that the statistical correlations between the viewing of violence and aggression found in longitudinal studies tend to be low. In any case, correlations in themselves do not reflect causal relationships. An effective design that would help to decide whether TV violence affects viewers' aggression would involve isolating a group of people, preventing them from watching TV, and exposing to heavy TV violence another group, comparable to the first one in all other respects. For ethical and practical reasons, such research can and must not be carried out. Therefore we are left with the possibility of studying the effects of viewing TV violence through 'follow-up' studies under naturalistic conditions or through the study of short-term effects in laboratory conditions. Since the late 1950s several thousands of such studies have been conducted. For comprehensive reviews see Eysenck & Nias (1980), Geen (1983) and Huesmann & Eron (1986). An integrative model for the different results is offered by Groebel (1986). The reviewers' overall conclusion is that the empirical evidence that filmed violence enhances aggression in the viewer is convincing. The studies on the effects of filmed violence can roughly be divided into laboratory and field research. By necessity, the laboratory research is concerned only with short-term effects, whereas field research attempts to study long-term effects and the role of television and film as factors in the environment of the viewer.

Laboratory research

When a company advertises a washing machine or a deodorant on the TV at high costs, it trusts that the viewers will imitate the behaviour of the actors on the commercial. To enhance the modeling effect, the young people who use the product are handsome and their background is as attractive as possible, with expensive cars, beautiful scenery, and music. The advertisers have understood some central psychological mechanisms of imitation: the model is imitated more readily if it has status, if it is attractive, and if the behaviour to be imitated seems to bring rewards or other beneficial consequences to the model.

In laboratory experiments, the same mechanisms have been found effective for the modeling of aggressive behaviour (Bandura, 1986). Children imitate the aggressive actions of a model portrayed in a film. Adults are imitated more than children, males more than females, and whites more than blacks. The model is also imitated more readily if he receives rewards than if he does not, or than if he is punished for his aggressive behaviour.

The effects are not restricted to the imitation of individual actions, for children have been observed to play more aggressively in general after looking at such films than after looking at neutral control films. Even verbal aggression has been shown to increase after the witnessing of physical violence on film. This indicates that viewing violence increases aggression in general, and not only the imitation of some specific acts. Imitation or modeling is the psychological process that has been studied most in the earlier research on the effects of film, but lately there has been a tendency to focus on broader effects of film and TV on the behaviour and personality of the viewer, and on the facilitating and inhibiting effects of personality and social factors on viewing.

Other examples of laboratory research involve studies of viewers' reactions while they are viewing violence. For instance, facial expressions (Lagerspetz & Engblom, 1979) and psychological reactions to violent scenes in the films have been observed, and/or the viewers have been interviewed about the specific scenes after viewing (Björkqvist & Lagerspetz, 1985). This can be used to analyse the film material through the viewers' reactions. It was found that children's physiological reactions to violent scenes were less intense when they were looking at the films with an adult than when they were alone. An inverse relationship between aggressive behaviour and anxiety was also demonstrated in experiments by Björkqvist. The films that evoked anxiety did not enhance aggressive reactions in the children's play after the viewing of the film. This relation is supported by the finding of Björkqvist & Lagerspetz (1985) that children with abundant aggressive phantasies became less anxious while watching violent scenes.

Field research

Most of the field studies have been concerned with long-term voluntary viewing. To study what people choose to view instead of manipulating the viewing in a balanced design does not, of course, fulfil the requirements of a good piece of research. If positive correlations between the viewing of violence and aggression are obtained, they may reflect the fact that aggressive people choose to view more violence, rather than indicating that viewing violence enhances aggression. Yet, to study voluntary viewing is the only possible way to approach the question of longitudinal exposure in a natural social setting.

In this type of research, the subjects are typically asked what programmes they have been looking at, the programmes are evaluated for their violent content, and the aggressiveness of the subjects is tested. The effects of viewing violence are then considered to be reflected in the correlations between amounts of violence seen, and aggressiveness of the viewers. To study long-term effects, the correlations between earlier viewing and later measures of aggressiveness are calculated.

Since aggressiveness is a multiply determined personality trait, and this is recognized by the researchers, the longitudinal research projects usually include a multitude of other data about the subjects. Such are data about the sex and age of the subjects, their TV viewing habits, their personality (especially aggressiveness), their relationships with their peers, and the social status of their family. Also considered are the parents' way of educating the subjects (for instance, use of corporal punishment, approving or rejecting attitudes towards the child, ways of interacting with the child, like reading stories), the TV viewing habits of the parents, their personality (especially aggressiveness), their evaluation of the child's viewing of TV (amount and quality), and other variables obtainable through interviews and tests.

All the above variables were studied in a longitudinal research project carried out in five countries: Australia (Sheehan), Finland (Lagerspetz & Viemerö), Israel (Bachrach), Poland (Fraczek), and the United States (Huesmann & Eron). The results were published in a book edited by Huesmann & Eron (1986). This project attempted to assess the invariance of the effects of the viewing of violence with identical methods in all the countries. Cultural comparisons have often been unreliable because different research methods were used in the participating countries. Even in the project in question, The Netherlands were originally included, but because the researchers chose to use somewhat different methods, their research was published separately (Wiegman, Kuttschreuter & Baarda, 1986).

Two cohorts (generations) of 7-year-old and 9-year-old children from all the participating countries were studied over 3 years. The aggressiveness of the children was measured through peer ratings and through self-evaluative questionnaires. In previous studies, peer rating has been found to be a very reliable measure of aggressiveness. Even its validity is likely to be superior to self-rating, and to observation, because it measures behaviour over a longer period and in more varied circumstances than can be covered by observing a class for a few hours. The class mates know each other from everyday interaction in the classroom and in the playground. They are able to point out with great consistency those of their peers who 'start a fight over nothing' or 'always make trouble'. Furthermore, if every pupil rates every other pupil, a large number of ratings can be obtained about each pupil in the class.

The viewing of TV violence was measured by asking the children of the different countries which television programmes they usually watched, which were their favourite programmes, and how often they looked at their favourites. The programmes were rated regarding their violent content. Identification with the heroes of the favourite series was measured by questionnaires asking how similar to these heroes the children felt themselves to be. The study was carried out during the late 1970s and early 1980s, when video equipment was still uncommon in most households in the countries in question. Since the children did not in principle have access to those films which are prohibited under certain age limits in cinemas, TV was the only, or almost the only, medium through which they could watch violent films. The correlations between TV violence and aggression were not very high but they were consistently positive for the different countries and for the different ages and genders. How are these statistical correlations to be interpreted? As we have seen, a positive correlation between the viewing of violent films and aggression can indicate that exposure to violent images makes the viewers aggressive, but it can also mean that aggressive persons choose to watch violent films, or that both are caused by some third factor or several other factors. There can also be a positive feedback loop, with the viewing of violence increasing aggression, and this, in turn, leading to an increase in the viewing of violence.

Several issues support the view that viewing violence has a causal effect on subsequent violence. First, as already stated, the correlations between average aggression and average TV viewing in our international study, although not very high, were consistently positive for different ages and sex groups. Secondly, consistently positive correlations were found in the different subgroups of the international study. More generally it has been estimated that around 75 per cent of field studies and 90 per cent of laboratory studies give a positive relationship between violence viewing and aggression.

Thirdly, a necessary though not sufficient condition that must be fulfilled before a correlation can be taken as reflecting a causal relation is that the cause must precede its effect. This is difficult to assess in field experiments, because the subjects are exposed to the viewing of violence throughout their whole lives. An exception was the study by Williams (1986) who studied communities before and after the introduction of TV and found increases of aggressive behaviour as a result.

However, it is possible to avoid the difficulty by comparing correlations between earlier viewing of violence and later aggression with the correlations between earlier aggression and later viewing of violence. In the international study in question, multiple regression analyses showed that in five countries later aggression was more predictable from earlier TV habits and earlier aggression than later TV viewing was predictable from earlier aggression and TV habits. In the Finnish material, first-year TV viewing and identification with TV heroes was the best predictor of aggression in the third year, thus exceeding as predictors the social status of the parents, their child rearing attitudes, and other variables involved.

The occurrence of some significant correlations between earlier aggression and the later viewing of violence indicates also a circular relationship or positive feedback loop with the viewing of violence and aggression enhancing each other.

Of course, this does not mean that the viewing of violence is the only causal factor for aggressive behaviour. Many other factors are important, and some of them operate by facilitating the effects of the media. The most important factor facilitating the effect of the viewing of violence on aggressiveness in the cross-national study was identification with the heroes in the film. It is remarkable that identification also with figures other than the aggressive ones correlated with aggression. This suggests that identification with any of the persons on the screen makes the viewers more likely to be influenced by any part of what they see. If the models are aggressive, as they are in many programmes nowadays, the result is an increase in aggressiveness. Sophisticated adults are, however, able to see all kinds of programmes with greater detachment than young persons.

Other studies have also shown that children who have difficulties in their social relationships, and do not have other adults to identify with, are especially likely to identify more with the figures on the screen. Children who have problems in their families have also been found to watch more TV. The viewing of violence is thus not a necessary nor a sufficient condition for aggression, but it leads to aggression only when other, specific conditions are also fulfilled. One task for future research in this area is to reveal these conditions with greater specificity than before.

A final issue is that the view that television violence can enhance aggressiveness is in harmony with many facts from research and from everyday life. Data from developmental psychology also point in the same direction. Children construct their picture of the world on the basis of what they see in the society around them. This experience includes the material offered by the mass media. If the mass media idealize violence and present it as an acceptable way of solving conflicts, and as a means to enhance one's own power and self esteem, then many young people will adopt the same attitudes and ways of acting. Also, psychological processes known to operate in other contexts could operate here too, for instance, imitation of the film actors, lessening of inhibitions against violence, desensitization to violence, and changing the norms of the viewers to be more favourable towards violence.

The discussion in this article exemplifies some of the difficulties encountered in psychological studies of problems that cannot be isolated from their social context. As mentioned in the introduction, the limitations of such research must be accepted and compensated for by frequent replications of the study in different contexts. This has been done in the research discussed here and the present evidence is clearly strong enough to warrant limitations on the supply of violent programmes in the media.

References and further readings

Bandura, A. (1986). *Social Foundations of Thought and Action. A Social Cognitive Theory*. Englewood Cliffs, New Jersey: Prentice Hall.

Björkqvist, K. & Lagerspetz, K. M. J. (1985). Children's experience of three types of cartoon at two age levels. *International Journal of Psychology*, **20** (1), 77–93.

Eysenck, H. J. & Nias, D. K. B. (1980). *Sex, Violence, and the Media*. London: Granada.

Freedman, J. L. (1984). Effect of television violence on aggressiveness. *Psychological Bulletin*, **96** (2), 227–46.

Geen, R. G. (1983). Aggression and television violence. In R. G. Geen & E. I. Donnerstein (eds), *Aggression. Theoretical and empirical Reviews, 2*. New York: Academic Press.

Groebel, J. (1986). International research on television violence: synopsis and critique. In L. R. Huesmann & L. D. Eron (eds), *Television and the Aggressive Child: a cross-national comparison*. Hillsdale, New Jersey: Lawrence Erlbaum Associates.

Huesmann, L. R. & Eron, L. D. (eds) (1986). *Television and the Aggressive Child: a cross-national comparison*. Hillsdale, New Jersey: Lawrence Erlbaum Associates.

Lagerspetz, K. M. J. (1984). Aggression. An instinct or not? In L. Pulkkinen & P. Lyytinen (eds), *Human Action and Personality. Jyväskylä Studies in Education, Psychology and Social Research*, **54**, 130–40. Jyväskylä: University of Jyväskylä.

Lagerspetz, K. M. J. & Engblom, P. (1979). Immediate reactions to TV-violence by Finnish pre-school children of different personality types. *Scandinavian Journal of Psychology*, **20**, 43–53.

Wiegman, O., Kuttschreuter, M. & Baarda, B. (1986). *Television Viewing related to Aggressive and Prosocial Behaviour*. Den Haag: Stichting Onderzoek van het Onderwijs.

Williams, T. M. (Ed.) (1986). *The Impact of Television. A Natural Experiment in Three Communities*. Orlando: Academic Press.

14

Cultural factors, biology and human aggression

MARSHALL H. SEGALL

Introduction

In the debate among students of human aggression between the extreme biological determinists and those who emphasize social learning, cross-cultural research may offer good prospects for clarifying both cultural and biological contributions to aggression. Data collected in a single society never permit empirical resolution of nature/nurture arguments, but cross-cultural psychology and anthropology have over the years contributed to a more balanced bio-social view by revealing both pan-cultural generalizations about aggression and some provocative cultural differences.[1]

From a cross-cultural perspective, aggressive behavior is learned from others, primarily through socialization and enculturation, involving both teaching and learning by observation. Social learning theories of aggression would find implicit support in cross-cultural data showing systematic variations in aggression, and when these variations are correlated with specifiable ecocultural differences, just *how* aggression is learned may be spelled out.

Aggressive behavior has multiple antecedents, many of which are ecological and structural. These include the probability of conflict over resources, the probability of frustrations of various kinds, norms governing conflict resolution, child-rearing emphases, and the kinds of behaviors displayed by persons who might serve as models to be emulated. All

---

[1] See Boyd & Richerson (1985) and Campbell (1975), for analogous features of (and differences between) biological evolution and cultural evolution. These perspectives deny the very meaningfulness of a nature vs nurture question but instead acknowledge the interaction of cultural variables with biological ones.

of these antecedents to aggression exist in every culture but they vary in kind across cultures. So, cross-cultural research on aggression is not only possible, it is essential.

Biology and aggression in cross-cultural perspective

As we have just asserted, aggressive behavior is culturally influenced, with this influence playing itself out largely through culturally mediated childhood experiences. Where, then, do we stand on possible genetic and other biological influences on aggression?

Our conceptual framework allows for the possiblity that some behavioral dispositions that are genetically based, or a product of human biological evolution, interact with experience-based dispositions to produce aggression as one of several reactions to frustration or as a conflict-resolution strategy. But these dispositions are not what are popularly thought of as 'instincts'.

That aggression is pervasive does not require that we consider it instinctive. Other biological and cultural forces could be common enough to contribute to aggression everywhere. Various social learning theories which stress the importance of socialization can also explain the pervasiveness of aggression.[2]

Nevertheless, we cannot and do not exclude biology from our effort to understand human aggression. There are many ways in which biology may be implicated. For example, there are possible dietary factors involved. Bolton, one of a growing number of anthropologists to stress biological factors in human behavior, has linked hypoglycemia (problems related to glucose metabolism) to homicide among the Qolla, an Amerindian group living in the Andes. Many anthropologists, however, says Bolton (1984), 'still think they can exclude biological factors from their attempts to understand social and cultural phenomena' (p. 2). We concur with Bolton, who asserted, 'Only by integrating social and cultural factors with psychological and biological ones will it be possible to solve some of the most significant anthropological puzzles' (1984, p. 2). Human aggression is surely one of those puzzles.

A second biological approach is to conceive of some aggressive behavior as a syndrome of traits thought to characterize persons who

---

[2] Social learning theories of aggression and the evidence that supports them have been presented in detail in Segall (1976), where a basic biological fact, the proto-aggression that characterizes infant behavior, is combined with social learning theories to account for the widespread occurrence of aggressive behavior.

perform the behavior, and then to enquire into the factors which are correlated with it. Taking this approach, Wilson & Herrnstein (1985) concluded from a review of over 1000 diverse studies of criminal behavior (done mostly in the industrialized part of the world) that genetic and bio-psychological variables are at least as important as sociological ones in predisposing individuals to become criminal. Specifically, they spotlight low intelligence and 'impulsivity' as predisposing factors, arguing, for example, that individuals of low intelligence are less aware of long-range consequences, less willing to defer gratification, and less able to restrict impulsivity, and, hence, more likely to tend toward criminality. The Wilson & Herrnstein argument is one version of a biological explanation of aggressive behavior in which intelligence and impulsivity are construed as genetically determined characteristics which in turn predispose individuals to commit criminal acts. But the mere fact that IQ scores and crime are correlated is, of course, subject to several possible explanations, many of which have little or nothing to do with genetics.

Another way in which biology might affect aggression is via androgen or testosterone levels. Reason to consider hormonal factors resides in the fact that, for non-human animals, castration (which dramatically lowers androgen levels) lowers aggressiveness but, at the human level, research results are contradictory; only some studies find a relationship between androgen level and aggression in humans. However, human studies of individual differences among adult males in circulating testosterone levels show correlations with (a) various physical characterstics and (b) acts of dominance. Both of these might in turn correlate with aggression. (Then again, they might not.) Whether they do or not may well depend on how those physical characteristics or tendencies to strive for dominance interact with cultural determinants of behavior. For example, in some societies, physically strong, dominating individuals may be rewarded for behaving aggressively. I will return to this particular set of biological factors later (cf. Herbert, chapter 6, this volume).

I include concern for biology in this conceptual framework, because it is reasonable to expect that certain biological factors, including but not exclusively genetic ones, will predispose individuals to react in particular ways to particular runs of experiences they might have in their lifetimes. But this makes it all the more important to conduct research on those experiential factors. The known biological facts regarding aggression cry out for elucidation through research on the cultural factors that mediate between biology and aggressive behavior.

One manifestation of aggression that has received some attention cross-culturally is crime. Criminal behavior includes many kinds of acts and reflects many different motives, including greed, genuine need, compensation for low self-esteem, and probably many others. But most acts that are socially defined as criminal involve the infliction of harm to others. Like other aggressive behaviors, criminal acts occur in all societies.

Landau (1984) examined statistics on criminal acts of violence in a dozen industrialized nation states for more than a decade spanning the mid-1960s through the late 1970s. During that period, all but one reported either stable or increasing homicide rates and robbery rates (Japan was the exception). Landau's primary use of these data was to provide a preliminary test of a model which predicts that the probability of violence and aggression as reactions to stress will increase when social support systems fail or malfunction. Using inflation rates as his major measure of social stress and ratio between marriage and divorce rates as his main social support measure, Landau found parallels in changes of these indices in every country but Japan.

There, an increase in social stress (as indexed by inflation) and a decrease in the strength of the family was accompanied by a considerable increase in suicide. As possible reasons for this departure from the overall picture found in this 12-nation sample, Landau suggested that strong social control mechanisms beyond the family exist in Japan, notably in schools, local communities and work places. He also cited the fact of much citizen participation in crime prevention in collaboration with professional law enforcement agents in Japan, as well as three important social control mechanisms embedded in Japanese cultural phenomena, namely a strong sense of shame, a sense of duty and loyalty, and respect for human relationships. Switzerland was another partial exception in this sample, in that while measures of aggression and violence increased, there was a greater increase in suicide rate.

Landau's data reveal some notable differences across countries in levels of crime rates. These cross-country differences, of course, must be interpreted very cautiously, since record-keeping varies so much between countries. Still, they are interesting. For example, some countries had relatively high and fairly consistently increasing homicide rates (Finland, Israel, USA and West Germany), while others had relatively low and stable ones (Austria, Switzerland, England, Netherlands, Sweden, Norway, Denmark and India) and Japan had both a low and declining homicide rate. So, even industrialized societies, all containing similar social stresses and experiencing declines in certain social support systems, vary in their crime rates. This suggests that other variables must be implicated in the story of cultural influences on crime.

The Landau study of crime concerned contemporary industrial nations. Bacon, Child & Barry (1963), beginning with a sample of 110 mostly preliterate societies, found 48 for whom ethnographic information in the Human Relations Area Files was adequate to permit reliable ratings of criminal behavior. They examined correlates of crime in general, correlates of theft, and correlates of crimes against persons. Their most striking finding was that both subcategories of crime – and hence crime in general – were more frequent in societies in which opportunity for contact between child and father is minimal (for example, societies with polygynous mother/child households).

The Bacon team related this finding to the tendency of most crimes to be committed by males and to findings concerning cross-sex identification problems in 'low-male-salience' societies. They offered their finding as support for a hypothesis that crime is partly a defense reaction against initial feminine identification in males. Such a hypothesis has been favored by several students of crime and delinquency in the United States. It is impressive to find supportive evidence for such a mechanism in a broad cross-cultural sample.

The two different forms of crime (theft and personal) were shown over a sample of 48 societies to have their unique correlates as well as a common relationship with factors likely to encourage compensatory efforts to establish masculine identity. The Bacon, Child & Barry study, then, yielded support for a social-learning interpretation of aggression, one that takes into account some possible consequences of gender.

Gender, social learning and aggression

In the preceding section of this chapter, we learned some facts from cross-cultural research about maleness and aggression and were introduced to the idea that masculine identity may play a role in aggressive behavior. In this section, we shall concentrate on material that relates to the social learning that leads to gender identity and to possible linkages between gender and aggression.

In her review of many cross-cultural studies dealing with sex differences, the anthropologist Ember (1981) designated aggression as the most consistent and best documented difference cross-culturally. Males are consistently more aggressive in the United States (Maccoby & Jacklin, 1974), in the several societies studied in the Six Cultures study

(Whiting & Whiting, 1975) and in 71 per cent of 101 societies for which ethnographic reports were examined by Rohner (1976).

With regard to criminal behavior, as noted above, Bacon, Child & Barry (1963) showed that in non-industrialized societies males commit the majority of criminal acts. In the United States, it has long been the case that the best predictor of fluctuations in crime rates is the proportion of the population composed of adolescent males. As A. P. Goldstein (1983) noted, 'the 1960–1975 increase in violent crime and the stabilization of the crime rate since 1975 parallel directly the number of 14–24 year-old males in the United States' (1983, p. 439).

So, there is a correlation between sex and aggression and a relationship between age and aggression. The perpetrators of most crimes in the United States are male adolescents. Similar findings prevail in Japan and other industrialized societies, as summarized in Newman's (1979) profile of the most typical violent individuals: 15 to 30-year-old males, with lower socio-economic status, living in urban areas, and disproportionately likely to be a member of an ethnic group that is low in the social hierarchy in the country. Naroll (1983) compiled 'juvenile criminal ratios' for 42 such societies (e.g. West Germany, New Zealand, Australia, etc.). In all of these societies, at least a quarter and as many as half of all reported crimes were committed by adolescent males.

Thus, we have cross-cultural evidence for a pan-species generalization about aggression. Males of the species perform most aggressive acts and they are most apt to do so as they move from childhood towards adulthood. What should we make of this?

A biosocial answer

Circulating testosterone is related to dominance behavior (Mazur, 1976). Mazur (1985) noted

> As young primate males pass through adolescence, they often become more assertive with posturing and strutting that may be labeled 'macho' in human terms ... they move rapidly up the group hierarchy, taking their place among the adult males. These changes may be a consequence of the massive increase in testosterone production that occurs during puberty.   (p. 383)

Since there is a surge of male testosterone at adolescence, that alone may produce an intensification of dominance–striving behavior among male adolescents and, if that behavior includes aggressive acts (which constitute one possible form of asserting dominance over others), then a sex-linked, age-related hormonal phenomenon could account for the high frequency of male adolescent aggression.

The validity of this answer depends on how closely dominance and aggression are linked. Mazur's (1976) review included some evidence that testosterone-produced differences in early development made males both more dominant and more aggressive, but Mazur (1985) concluded that dominance and aggression are not inextricably linked (see Herbert, this volume). In the course of presenting a model of competition for status, which emphasizes long-term changes in testosterone, Mazur (1985) noted that the literature links testosterone and dominance behavior and that it is important to distinguish dominance behavior from aggressive behavior especially for humans, 'who often assert their dominance without any intent to cause injury' (p. 382).[3] Whether dominance-striving by male adolescents includes aggressive acts probably depends, in the end, on cultural norms.

A strictly cultural answer

There would be a very simple answer to the question of why most aggressive acts are committed by male adolescents (an answer that leaves out biology altogether) if cross-cultural research showed that in most societies boys are encouraged more than girls to behave aggressively. Then we could say that male adolescents are more aggressive simply because they have been taught to be. But this purely cultural answer is too simple.

Barry *et al.* (1976) scored nearly 150 societies drawn from the Standard Cross-cultural Sample (Murdock & White, 1969) on inculcation of aggression among children. They found a sex difference on the average over all of these societies, but the sex difference was significant in only one out of five of these societies when they were examined singly. So the cross-cultural consistency in greater male aggressive behavior cannot be attributed solely to differences in inculcation of aggression.

Other factors must be implicated in the phenomenon of greater aggressive behaviors among males than merely their hormones or the fact that generally they are subjected to more inculcation of aggression than are females.

[3] Mazur also notes that at present there are no firm data on the effect of testosterone on dominance in humans. Also, the causal link between testosterone and dominance behavior may be in the opposite direction, with success in status competition producing increase in testosterone. He therefore refers to the relationship between testosterone and dominance behavior as 'reciprocal' (1985, p. 383). (This issue is discussed in more detail by Herbert, this volume: eds.)

An expanded bio-cultural model

Whatever biological mechanisms are involved, they probably interact with cultural mechanisms. These must be more complex than inculcation alone. They include division of labor by sex, gender identity, and aggressive behavior that serves a gender-marking function.

Division of labor by sex.   Every society has some division of labor by sex and some linked modal sex differences in behavior. Most pertinent is the notion that the division of labor by sex sets the stage for differentiation between the sexes in socialization emphases and that this differentiation in turn functions as a means for preparing children to assume their sex-linked adult roles. Thus, the relationship between division of labor by sex and differentiation across the sexes in socialization emphases is a reciprocal one.

The clearest of all (and most nearly universal) sex-linked adult roles is child-rearing itself. During socialization, females are taught traits that are compatible with child-rearing and are later encouraged to assume that role. Males, on the other hand, are taught other traits during childhood, like independence, and encouraged later to assume roles (e.g. food-getting) that are largely incompatible with child-rearing. Consequently, females do most of the child-rearing in most societies and virtually all of it in some.

Whatever the reasons for this, there results, in many societies – and perhaps in all in varying degree – a paradoxical state of affairs. Young males have somewhat restricted opportunity to observe adult males since their fathers tend to be non-participants in the child rearing. To the extent that gender-role learning involves modeling (or learning by observation), boys will have restricted opportunity to acquire a masculine identity early in life.

Cross-gender identity.   The cutting edge of this paradox is that father-absence is most marked in societies with the sharpest division of labor by sex. Thus, precisely in those societies in which the two sexes are expected to have the most distinct gender identities, young males have restricted opportunity to acquire their masculine identity by emulation of male models!

In societies with a particularly distinct division of labor by sex,[4] there

---

[4] Division of labor by sex is likely to be sharper in sedentary societies, in those which cultivate large grain crops, and in those which raise large animals.

is therefore a likelihood that young males will acquire a cross-sex identity. How ironic this is, considering that in societies where there is a relatively sharp gender-role distinctiveness, the role of women is often regarded with contempt by males, whose own activities are accorded higher prestige. Consider the pressure that adult males in such societies must face to avoid behaving 'in womanly fashion'. Yet their sons, we are here suggesting, are likely, during childhood, to acquire a predominantly female identity.

Gender-marking aggression. The Bacon *et al.* (1963) study of crime was cited above as a primary source of evidence for the masculinity of aggression, showing, as it did, that males commit the preponderance of crimes in most societies. This same study also revealed that aggressive crimes, such as assaults, rapes and murders, were more likely to occur in societies that provided exclusive mother–child sleeping arrangements, which prevail, of course, in societies where fathers are not active participants in child-rearing. Bacon *et al.* related this finding to the ideas of 'cross-sex identity' that had been introduced earlier as a likely problem to be found in 'low-male-salience' societies. Applying this idea to crime, Bacon *et al.* offered the hypothesis that aggressive crimes are part of a defense reaction against initial feminine identification in males.

Extending the Bacon *et al.* hypothesis, the anthropologist Whiting (1965) suggested that males reared primarily by females would be more susceptible to envy of powerful adult males, but could not become like them until escaping somehow from the early influence of their mothers.

Whiting (1965) linked the young males' status envy to what she called 'protest masculinity'. Observantly she suggested,

> It would seem as if there were a never-ending cycle. The separation of the sexes leads to a conflict of identity of the boy children, to unconscious fear of being feminine, . . . exaggeration of the difference between men and women, antagonism against and fear of women, male solidarity and hence to isolation of women and very young children. (p. 137)

The identity conflict designated by Whiting would obviously constitute a problem for societies that encourage sharply distinguished gender roles. In some such societies, the problem is dealt with in an institutionalized manner, viz. by male initiation ceremonies. Severe male initiation ceremonies at puberty, often including tests of endurance and manliness, were found by Whiting, Kluckhohn, & Anthony (1958) to be correlated with exclusive mother–son sleeping arrangements and post-partum sex taboos, both indices of father absence. The interpretation of this finding that is pertinent to our present discussion is that such ceremonies serve the function of stamping in masculinity for boys who need it due to inadequate opportunity to acquire it in childhood.

What happens, however, in societies which have this identity conflict but lack the initiation ceremony?

In such societies, which have the preconditions requiring a stamping-in of masculinity – but which don't achieve this via initiation ceremonies or other institutionalized practices – adolescent males will try on their own to assert their masculinity. They may do so in a variety of ways but one of them might well be to behave aggressively. If a society is one in which aggressiveness and such allied traits as fortitude and courage are an integral part of the definition of manliness, boys approaching manhood will wish to display these characteristics.

They may have been taught and encouraged to behave in this way but in addition to whatever inculcation may have occurred during childhood, a structural feature of such a society can set the stage for the boys' need to acquire these behaviors and traits during adolescence. This structural feature is a sharply defined division of labor by sex, with child-rearing assigned primarily or even exclusively to mothers, resulting in relative father absence during the boys's childhood. That in turn leads to cross-sex identity during childhood that has to be undone by displays of the 'manly' behaviors and traits, notably fortitude, courage and aggression. Such aggression has the function of displaying that the actor is behaving like an adult male, in accord with that society's definition of the masculine gender.

Cultural variations in aggression-modeling

Those societies which provide more aggressive models should, according to a social learning theoretical approach, encourage more aggression.[5] This proposition has received much study, particularly in the United States, where many social scientists have addressed the question of the impact of violence in the media, especially films and television (see chapter by Lagerspetz). The prevailing view among social

---

[5] If the availability of aggressive models increases aggression, why did we argue earlier that the *absence* of adult males can increase aggression in their sons? The father absence of which we spoke earlier occurs during early childhood; its effect is to set the stage for later learning of aggression, by various means including, as we are now stressing, the emulation of aggressive models.

scientists who have studied the effects of the massive diet of violence that exists in the American media is that those viewers who are most prone to be aggressive (such as the young males we have described above) find encouragement and tutelage from television and films. Those who control the media in that culture, the television network executives and the film producers, tend to resist the conclusion that the media make a causal contribution to the problem of aggression in America, and wherever else American films and TV programs appear, but that conclusion is hard to avoid.

Groebel (1986) has reviewed studies done in other societies where media effects on aggression have been examined, including a cross-cultural analysis of five different countries initiated by Huesmann & Eron (1986). In a study in Australia, where government officials monitor TV programs in an effort to minimize violent content, Sheehan found evidence of overall smaller effects than typically found in US studies, but boys who identified with aggressive male TV characters were those who themselves displayed aggressive behavior. In Finland, where no locally produced programs contain any violence, but imported US programs do, viewers tend to perceive what violence they see as characteristic of the US but not of Finland and therefore are less likely to emulate it. But again, Groebel, and Lagerspetz & Viemerö (1986), find evidence in some Finnish data that boys who identify with male TV models find encouragement to develop their own aggressiveness. Groebel also cited data from Israel, Poland, the Netherlands and West Germany and concluded that in all of these societies the impact of media violence on real aggressive behavior is less than in the United States because other factors (e.g. cultural norms, parental training) interact with the media violence to produce varying degrees of inhibition of aggression.

While media violence can contribute to aggression in any society, the extent to which it does so appears to depend on the cultural context in which the media violence is made available. Media violence interacts with other cultural variables to produce a complex effect. To understand its effect, additional cross-cultural research is clearly needed.

Conclusion

This chapter offers a framework and a rationale for studying aggression cross-culturally. I have argued that we cannot understand human aggression without viewing it from a cross-cultural perspective. While biology is surely implicated, it interacts with our culturally shaped experiences to lead us to react to frustration, to assert dominance, and to attempt to resolve conflicts in a diversity of ways. While males out-aggress females probably everywhere in the world, even this biologically rooted phenomenon cannot be understood without taking into account cultural factors.

References and further reading

Bacon, M. K., Child, I. L. & Barry, H. III. (1963). A cross-cultural study of correlates of crime. *Journal of Abnormal and Social Psychology*, **66**, 291–300.

Barry, H. III., Josephon, L., Lauer, E. & Marshall, C. (1976). Traits inculcated in childhood: cross-cultural codes V. *Ethnology*, **15**, 83–114.

Bolton, R. (1984). The hypoglycemia-aggression hypothesis: debate versus research. *Current Anthropology*, **25**, 1–53.

Boyd, R. & Richerson, P. J. (1985). *Culture and the Evolutionary Process*. Chicago: University of Chicago Press.

Campbell, D. T. (1975). On the conflicts between biological and social evolution and between psychology and moral tradition. *American Psychologist*, **30**, 1103–26.

Ember, C. R. (1981). A cross-cultural perspective on sex differences. In R. H. Munroe, R. L. Munroe & B. B. Whiting (eds), *Handbook of Cross-Cultural Human Development*. New York: Garland.

Goldstein, A. P. (1983). United States: causes controls, and alternatives to aggression. In A. P. Goldstein & M. H. Segall (eds), *Aggression in Global Perspective*. Elmsford, New York: Pergamon Press.

Groebel, J. (1986). International research on television violence: synopsis and critique. In L. R. Huesmann & L. D. Eron (eds), *Television and the Aggressive Child: a cross-national comparison*. Hillsdale, New Jersey: Erlbaum.

Huesmann, L. R. & Eron, L. D. (eds) (1986) *Television and the Aggressive Child: a cross-national comparison*. Hillsdale, New Jersey: Erlbaum.

Lagerspetz, K. & Viemerö (1986) Television and aggressive behavior among Finnish children. In L. R. Huesmann and L. D. Eron (eds.), *Television and the Aggressive Child*. Hillsdale, New Jersey: Erlbaum.

Landau, S. F. (1984). Trends in violence and aggression: a cross-cultural analysis. *International Journal of Comparative Sociology*, **24**, 133–58.

Maccoby, E. E. & Jacklin, C. N. (1974). *The Psychology of Sex Differences*. Stanford: Stanford University Press.

Mazur, A. (1976). Effects of testosterone on status in primary groups. *Folia Primatologica*, **26**, 214–26.

Mazur, A. (1985). A biosocial model of status in face-to-face primate groups. *Social Forces*, **64**, 377–402.

Murdock, G. P. & White, R. R. (1969). Standard cross-cultural sample. *Ethnology*, **8**, 329–69.

Naroll, R. (1983). *The Moral Order: an introduction to the human situation*. Beverley Hills: Sage.

Newman, G. (1979). *Understanding Violence*. New York: Lippincott.

Rohner, R. P. (1976). Sex differences in aggression: phylogenetic and enculturation perspectives. *Ethos*, **4**, 57–72.

Segall, M. H. (1976). *Human Behavior and Public Policy: a political psychology*. Elmsford, New York: Pergamon Press.

Sheehan, P. W. (1986). Television viewing and its relation to aggression among children in Australia. In L. R. Huesmann & L. D. Eron (eds), *Television and the Aggressive Child: a cross-national comparison.*. Hillsdale, New Jersey: Erlbaum.

Whiting, B. B. (1965). Sex identity conflict and physical violence: a comparative study. *American Anthropologist*, **67**, 123–40.

Whiting, B. B. & Whiting, J. W. M. (1975). *Children of Six Cultures: a psycho-cultural analysis*. Cambridge, Mass.: Harvard University Press.

Whiting, J. W. M., Kluckhohn, R. & Anthony, A. (1958). The function of male initiation ceremonies at puberty. In E. E. Maccoby, T. Newcomb & E. L. Hartley (eds), *Readings in Social Psychology* (third edition). New York: Holt.

Wilson, J. Q. & Herrnstein, R. J. (1985). *Crime and Human Nature*. New York: Simon & Schuster.

E. THE MACRO LEVEL: SOCIETIES AND NATIONS

Editorial

This section deals primarily with a macro phenomenon, war. Some of the principles described in the previous sections can be applied also to the macro level. Political and military action is taken by individual leaders who share physiological and psychological characteristics with other human beings and are socialized along similar lines. Decisions are made in groups with dynamics and communication patterns similar to those described in the preceding chapters. And yet, these factors are not sufficient to explain aggression on the macro level, that is *war*. Some of them may even be absent. Individual aggression probably plays little part in modern war, where 'rational' data and information processing is more important than personal courage and high-scale aggressiveness, where cooperation and support in the preparation of technological strategies are more important group characteristics than aggressive interactions or the need for an individual scapegoat. Even the individual soldier is motivated more by considerations of obedience, loyalty and/or fear than by aggression.

For such reasons, simple analogies cannot be used across the different levels from intra-individual processes to the macro level to describe their respective phenomenology or to explain their causes and consequences. Each level involves increasing complexity and an interweaving of many different factors which may even include chance events. Individual charismatic leaders, economic interests at a given moment, and powerful groups can facilitate or trigger the *outbreak* of collective violence and war, but it usually demands long-term technical preparation and an historically and politically suitable climate to manage a war in a technologically developed area of the world.

Although direct analogies between the different levels may not be useful, a physical fight between two individuals and modern war differing in so many respects, the levels are *interlocked* in a reciprocal way: biological, psychological and social processes interact and are the stuff of which human development is made. Individuals constitute groups which in turn are the elements of a whole society. In a reciprocal sense, the society co-determines the socialization of the individual and the characteristics of the group. Thus, the interlinking between the different levels can be described in terms of a continuing reciprocal process, where each level consists of elements of the lower levels but involves additional properties arising from the complexity of the interactions involved (*see* Chapter 1, this volume).

Modern war must be regarded as a highly complex system and as an institution where different groups and individuals interact with each other, for instance government with staff, leaders with soldiers, and also each of these with technological and communication systems, such as computers and information pools. The efficiency of this institution is probably highly dependent on a common value or belief system, shared by all or most of the members of the institution. Reduction in the *perceived* complexity of the institution, simplification of the characteristics of the enemy, and description of the goals of the war in simple terms may help to create such a common belief system. This is one of the reasons why propaganda and psychological warfare play a major role in modern international conflicts.

One means to simplify the complexity is to use symbols and analogies at the individual level. Individual charismatic leaders or heroes may symbolize the values of the institution so that individuals can identify with them. If a common value system is absent, its place may be taken by an individual symbolic figure such as a strong leader. Another way to create a common goal or belief system is to focus people's attention on a simplified image of the enemy. Often the opposing party in a war is addressed as a single person, 'The Russian', 'The American', 'The German'. An individual can more easily be described, is a more concrete goal, and can more easily be given certain stereotyping attributes such as evil, arrogant and so on.

Thus, there are two parallel structures: one is the *actual* structure of war involving the different roles, interactions and processes in all their complexity. This complexity, involving dependence on complicated technological systems, makes it hard to predict the outcome of a global and even of a regionally limited conflict. The other structure is the *perceived* structure. As we have seen, this may involve a simplified and even an incorrect picture, or it may be nearly accurate, but it hardly ever consists of a complete representation of the actual system. It would seem to be especially important for decision makers to develop a view of the system which is near adequate, but as discussed above, group processes mitigate against this.[1]

For the perception of the system by those individuals or groups who, though not decision-makers, are nevertheless crucial parts of the interaction processes, information and media channels are important. They can contribute either to a simplification of the perceived structure or they can increase the likelihood that the system is perceived accurately and in all its complexity. The media are probably responsible for the fact that in many countries war is now perceived as unpredictable, as a risky game, by the majority of the population.

One of the goals of scientific analysis is to offer a representation of reality, including the institution of war, as adequate as possible. This may be possible on the micro level but is extremely difficult on the macro level. The problem stems from the fact that, unlike the observation and measurement of biological processes or even simple group interactions, no measures or empirical observations can be powerful enough to take account of every possible element in a complex system like war because of the multiplicity of interacting factors. In these circumstances, some authors focus on issues that they see as of overriding importance: for instance, Hinsley (in Väyrynen, 1987) argues that, in recent centuries, European wars have become *less* frequent as weapons have become more powerful because states have become more hesitant to go to war. (Interestingly, in the same volume, Haas discusses the relation between state interdependence and the incidence of wars and concludes that 'The world has got *more violent* as interdependence has increased' (our italics).) The authors of the two chapters in this section, however, emphasize that the factors leading to the outbreak of war are always multiple and complex. Winter uses the method of case histories, offering an historical analysis of World War I and World War II. Singer adopts a quite different approach, looking for regularities in the occurrence of war over time and attempting to relate occurrence to circumstances. Both authors come to the conclusion that there is no empirical evidence at all for the existence of a natural law for the ocurrence of war.

A multitude of factors in different combinations, many of them related

---

[1] Actually, it is not the 'bad intentions' of individuals or groups which seem to be the major risk factor on modern war but rather the belief that the complex system of war and its outcomes are completely controllable or predictable.

to information processing and decision making by an élite, can determine the outbreak of war. Omitting or changing one or a few of these factors can lead to a change in the whole system and thus to a change in the likelihood of war. Most of the factors can be classified into three groups: social factors, including their interactions as discussed above; temporal factors, including influences from periods preceding the actual events, the historical background, the process characteristics, and possible periodicities; and so-called 'spatial' factors, namely the environmental, societal and cultural factors in which the conflict is embedded. Of course, all three categories are interlocked: interactions develop over time, social interaction and communication constitute culture.

The time dimension can be used to demonstrate that war is by no means a regular phenomenon. There have been periods in history where the people of some areas of the world were frequently engaged in highly violent conflicts: for instance, during the 30-years' War the Swedish were regularly involved in aggressive and cruel attacks. Yet during other periods Sweden has been one of the least aggressive countries. Examples like these demonstrate both that countries are often not consistent in their readiness for war and that wars do not follow a regular periodicity: history does not repeat itself. At least, it never repeats itself if one considers a sufficiently differentiated, in-depth analysis of the respective events' phenomenology. Some, usually many, details will always be different, and this may change the whole system's characteristics. An adequate analysis would demand the determination of the specific combinations, the weights of the individual factors, and the possible effects of aggregating different factors.

The same principle holds for space. Comparing two or more similar cultures demonstrates that there is no necessary similarity in the like-lihood of war even under nearly identical conditions. The presence or absence of single aspects, like a leader or a specific missile system, may account for the difference.

All in all, aggression on the macro level, such as modern war, must be regarded as involving a highly complex causal network with elements as diverse as individual action, group decision making, existing and changing group values, present and future interests, random events, technical instruments of war, communication systems, and changes in all of these over time. Every violent conflict involves a different pattern of inter-actions between these factors. Sometimes there may be a considerable overlap between the causal networks operating in apparently compara-ble conflicts, but more often, at least at second sight, there are hardly any similarities. Some of the factors involved and some of their interactions, like group processes and the historical background, have already been systematically analysed. Others still demand analysis.

The fact that until now no automatic regularity or periodicity in wars has been identified falsifies the belief that war can be described in monocausal terms. This means that the assumption of a natural inevitability of war must be rejected.

Further reading

Abdel-Rahman, I. H. *et al.* (1986). *Disarmament and Development.* Declaration by the Panel of Eminent Personalities. New York: United Nations.

Bronfenbrenner, U. (1961). The mirror image in Soviet–American relations. *Journal social Issues*, 17, 45–56.

Gurr, T. R. (ed.) (1980). *Handbook of Political Conflict.* New York: Free Press.

Richardson, L. F. (1960). *Statistics of Deadly Quarrels.* Pittsburgh: Boxwood.

Singer, J. D. & Wallace, M. D. (eds) (1979). *To Augur Well: Early warning indicators in world politics.* Beverly Hills: Sage.

Väyrynen, R. (ed.) (1987). *The Quest for Peace.* London: Sage.

Neil E. Miller of the Yale University has supplied us with some classical references to cultures without war:

Henry, J. (1941). *Jungle People.* New York: J. J. Augustin Press.

Murdock, G. P. (1934). *Our Primitive Contemporaries.* New York: Macmillan.

Spencer W. & Gillen, F. (1927). *The Arunta.* London: Macmillan.

15

Causes of war

J. M. WINTER

The use of biological metaphors in historical study is a time-honoured and widely practised art. It is also extremely pernicious. There can be no better illustration of the dangers of this mode of thinking than in the field of 'war studies', or, as the French have termed it, *polémologie*. This school of thought, which flourished after 1945 under the auspices of the French government, sought to detect the causes of and nature of war much as a biologist would detect the causes and nature of disease. In wartime and post-war France, such views were commonplace, and may have deflected attention from the very unbiological submission of many Frenchmen to the Nazis during the Vichy regime of 1940–44.

What often took the form of colloquial or academic discourse was elevated into great literature by Albert Camus, whose masterpiece, *The Plague*, was written during the war, but published after the armistice. Less a parable about war than about totalitarian temptations, Camus' novel captured the dense claustrophobic atmosphere of a society unable to resist an unseen, but deadly enemy. In the hands of a genius, this literary vision is terrifying and morally powerful. But in the hands of lesser men and women, this kind of biological metaphor has frequently taken on the appearance of biological 'truth'.

If war were a biological fact, then the techniques of biological analysis would be applicable to it. This is the underlying assumption of a vast outpouring of studies on war and peace which have appeared over the past 20 years. Just like biologists or physicians, these polemologists – as the French variety call themselves, borrowing *polemos* from Heraclitus to dignify their specialty – define, dissect, compare and, wherever possible, measure the life histories of human conflicts. The results have not been uninteresting. We have statistical series of frequencies of eruptions of violence, and complex correlations between war and a host of other variables. We have chronicles of world violence, charting the spread of the 'disease' over time and space. We have studies of population pressures and geographical density as preconditions of war. We have attempts to compare the life and death of societies in war to the life and death of individuals.

In effect, we have a host of academic work which rests on the assumption that there is something either in human nature or in the nature of collectives which makes warfare – like illness, industrialization or fertility – an ineradicable part of history. Just as data on historical epidemics, economic development or population change are studied on the macro- as well as on the micro-level, so data on war must be collected, systematized and analysed. Just as the study of past outbreaks of disease, or of economic and demographic cycles, have entered into medical, economic and demographic forecasting, so polemology has taken on a predictive as well as a descriptive role.

It is the purpose of this brief paper to suggest that this approach is as futile as it is dangerous. It is a classic, though late, example of the positivist mind at work. That is to say, it replicates the naive, nineteenth-century optimism of August Comte and his followers, who believed that there are laws of history, that these laws are scientific, and that they can be discovered by a careful scrutiny and collection of the relevant evidence.

Positivism is the enemy of open historical research, because it presumes that a pattern must exist, and then finds evidence to justify the presumption. If we consider just two instances of historical causality, of some importance in contemporary debate about war and peace, we can see the futility of such an approach.

The instances I have in mind are the outbreaks of the First and the Second World Wars. My argument is that these events cannot be understood as part of a model of warfare or conflict-propensity which is abstracted from the specific, very human and absolutely particular circumstances of the time.

1914: the war of illusions

Let us consider first the celebrated case of the outbreak of the First World War. This historical episode is of interest first for intrinsic reasons: it launched a period of conflict which future generations may well term the second Thirty Years' War. But secondly, the slide into war in 1914 has been taken as analogous to the precariousness of the international balance of power of our own day. This analogy, in my view, is a mistaken one in terms of the nature of political conflict, then and now. But it is alarmingly appropriate when we analyse the ways in which the assumptions of the key actors in international diplomacy determined the nature and outcome of their policies.

The war crises of 1914 and 1939 are two classic cases of the outbreak of war being *primarily* a function of what was in the minds of the men who made the key decisions. It is not that economic, demographic or social considerations were irrelevant to their calculations; on the contrary. It is rather that the fateful steps they took to go to war in 1914 and in 1939 cannot be separated from the mental landscape they inhabited. The crucial importance of what one scholar has called the 'unspoken assumptions' of the men of 1914 is what makes it impossible to analyse this set of events, as it were, 'polemologically'.

The war crisis: what happened in 1914?

The key question about 1914 is how a political crisis in eastern Europe turned into the First World War. The basic chronology is not in dispute. The Austro-Hungarian Crown Prince and his wife were assassinated on a visit to Sarajevo in Bosnia on 28 June 1914. The Austrians blamed Serbian intelligence for this murder, which was carried out by Serbian students. An ultimatum was delivered to Serbia by Austria on 23 July.

This turned an Austro-Serbian quarrel into an Austro-Russian one, because any humiliation of independent Serbia touched Russian pretentions to defend all Slavic peoples. The result was the beginning of Russian military mobilization.

This turned an Austro-Russian quarrel into a Russo-German one, since Russian mobilization threatened Germany with a two-front war. This was because French hostility to Germany dated from the French defeat in the Franco-Prussian war of 1870–1, and because a Franco-Russian understanding on mutual defence had existed since the 1890s. As the Kaiser never ceased to say, Germany was thus 'surrounded' by enemies. Russian mobilization therefore led to German mobilization on 31 July 1914.

German mobilization was linked to a military plan which turned a Russo-German conflict into the Great War. In this plan – the Schlieffen plan – Germany avoided a protracted two-front war by knocking out the French army before Russian mobilization was completed. This was to be done by a huge flanking operation under which the German army would pass through Belgium and northern France in an enormous arc, and sweep up the outflanked French army near Paris, thus replicating the decisive victory of 1870.

The political costs of this plan were clear: it brought Britain into the war, as guarantor of Belgian sovereignty, and completed the spiral of conflict which had started with the murder of two Austrian nobles and ended with the deaths of nine million men.

Why war in 1914?

Thus far all seems clear. By and large, historians know what happened. What they have been arguing about virtually since the war ended was who was to blame for the escalation of a Balkan quarrel into a world war. Basically, there are three schools of thought on the question as to how to apportion blame for the outbreak of war and for the human catastrophe that followed. They are the 'primacy of German responsibility' approach, the 'calculated risk' approach, and the 'collective guilt' approach.

In a sense, all three contain part of the truth. Those who subscribe to the first approach rightly insist that at the heart of the war crisis was the German military. Given the nature of the German political system, the chancellor, Bethmann-Hollweg, could not possibly prevent war once the military and the Kaiser had made up their minds. This they had done by late July 1914, thereby making war inevitable. Primary responsibility for war must, therefore, rest in Berlin and nowhere else.

But it is likely that some of those at the centre of power in Berlin – including the Chancellor, Bethmann-Hollweg – were guilty not of provoking war, but rather of playing a game of 'brinksmanship' which went terribly wrong. This is the force of the second line of interpretation. In other words, Germany's leaders played a very dangerous game in backing the Austrian case against Serbia; this simply got out of hand. The war was thus not an outcome of long-term strategy, but of short-term miscalculations of the responses of the other side. War was a monumental failure of foreign policy, not its realization.

This too contains part of the truth. But other scholars have correctly moved beyond the field of German foreign policy to suggest that a more universal set of miscalculations and mistakes contributed both to the atmosphere in which the war crisis developed and to its ultimate resolution in war. In this school of thought, German belligerence is not excused but rather assumes an integral part in a wider European failure to keep the peace.

This argument accepts that the growth of German power made international politics inherently unstable in the pre-1914 period. But it underlines the essential point that structures do not go to war; individual states do on the advice of their leaders. Hence, to understand the causes of war on this most concrete level, we must appreciate how international affairs were perceived by key political figures in all the major powers. All bear responsibility for the war, because so many of the key decisions they made in 1914 were based on misperceptions, miscalculations and illusions.

Of these illusions, five stand out. The first was the illusion of the fail-safe and rational character of diplomacy. The second was the illusion that since war involved domestic risks, both political and economic, relatively unstable regimes would avoid it. The third was that war would be relatively short, a matter of weeks and months, not years. The fourth was that war could be controlled and limited. And the fifth was a view of war as rebirth, rejuvenation, renewal.

Every one of these key assumptions was profoundly and tragically without foundation. Even those who emphasize German war guilt or the 'calculated risk' argument share the view that war was a product of shared ideas, widely disseminated in pre-1914 Europe. Many of these notions were divorced from reality, no doubt, but they were none the less powerful for that.

For our purposes, this all too brief summary of a mountain of historical literature supports two general points. The first is that whatever the structural weakness of the European state system, or the particular, conjunctural features of the Balkan crisis, we cannot make sense of the outbreak of war without descending to the level of the individual. And, secondly, to understand why states go to war, we must enter into the mental world of those who lead them.

1939: Hitler's war

The same points may be made even more emphatically with respect to the outbreak of the Second World War. There is little need to go into the chronology of events leading to the German invasion of Poland on 1 September 1939, and the declarations of war between Germany on the one hand and Britain, France and their allies on the other. Historians agree on what happened, and unlike their colleagues who work on 1914, they also largely agree on why.

Nazi Germany under Adolf Hitler went to war as part of a plan. Perhaps the plan was to confront Germany's enemies a few years later, when full economic preparations had been made. But even if the timing was flexible, the design was not. When it came to the moment of decision, we clearly see the outcome of the ideas Hitler had developed openly since the 1920s. These proclaimed the need for Germany to reverse the humiliating terms of the peace treaty which ended the First World War and once and for all to break out of her confined position in central Europe by a thrust east and west.

When put in these terms – which are those contemporaries on both sides understood at the time – the issue of the causes of war becomes once again a matter of mentalities. Of course, Hitler operated within a given set of political, economic and social structures. No one can understand the Nazis' rabid hatred of communists and Jews without appreciating the special path of German state-building, accomplished precisely at the period of rapid industrialization, and thereby ensuring the fusion of political with social and economic tensions and frustrations.

But what mattered most was the specific ideological form in which these tensions were expressed. The kind of nationalism Hitler espoused was highly symbolic and celebrated the escape from rationality and the apotheosis of the Aryan race. But we should never forget that the negation of reason is a highly rational position, and in this case, one which was very cleverly exploited by a large number of highly intelligent people. They knew what they were doing, both when they created the environment in which war was glorified, and when they forced a set of very reluctant Western statesmen to return to the field of battle in 1939.

The general and the particular in the study of the causes of war

This rough sketch of the nature of two – and only two – war crises and their resolution in the twentieth century raises central issues about the appropriateness of biological models in the study of human conflict. The key difficulty lies in the abstraction of a set of variables from the events described above, in order to facilitate comparisons with similar data on warfare in other societies at other times.

Of course, data on trade cycle fluctuations, price levels, industrial conflicts, even on climatic conditions, throw light on the general environment of 1914, 1939, or of any other period in which questions of war and peace are debated and resolved. But these data in and of themselves are incapable of providing an explanation of the events in question. The key linkages are missing, and thus the connections remain abstract and general rather than specific and particular.

The particularity of war crises arises out of the particularity of ideologies, broadly defined as the mental furniture of a period. This is ultimately what makes the biological analogy in political history unacceptable. No one would deny that individual statesmen operate in a harshly constrained universe. Their freedom of action is certainly limited, by economic, geographical, political and social considerations. But on occasion, individual leaders can and do break through these constraints and establish a new reality based on ideas of transformation which become dynamic historical forces. This happened at least twice during this century: in the Russian Revolution of 1917 and in the Nazi seizure of power in 1933.

Scholars have termed such events overdetermined. In other words, the social forces pointing towards them were so varied and powerful that we can 'explain' them many times over. But the difficulty here is that such explanations are bound to be lopsided or incomplete if they fail to incorporate one key element; the propensity of social formations or coteries to seize the moment to convert their ideas into reality.

These ideas are certainly not fixed, and frequently change with alarming rapidity. Several light years separated conceptions of war in 1914 from those held in 1916. This we know from history. But of perhaps greater importance in this field is the recognition that a positivist approach to war tends to induce in those who subscribe to it a degree of pessimistic complacency about the possibilities of avoiding war or of eliminating it.

The polemologist would have it that certain regularities – *ricorsi* in the language of the philosopher Vico – in human behaviour exist, and among them is the propensity to go to war. Data about wars can therefore be collected, analysed and turned into historical laws. This on the face of it is either banal or misleading, for it has little chance of helping us to understand discrete war crises, which, after all, are the ones in which ordinary human beings live.

But what is not so harmless is the use of so-called historical laws, derived from the collation of evidence about the past, as predictive tools. No one could object to the statement, 'what was can tell us something about what will be'. But it becomes unacceptable in the form, 'what was is built into human nature and describes an immutable force against which it is futile to struggle'.

The polymological fallacy is, therefore, the presumption that 'war' is *intrinsic* to human society. But such a position is just as indefensible as the argument that since slavery was a distinctive feature of 5000 years of human history, the institution of slavery is 'natural'. When ideas about human dignity changed, so did the supposed immutability of slavery.

The same is true with respect to war. I have deliberately emphasized the importance of ideas in war crises, not only because they have been demonstrably central in these episodes, but also because the realm of ideas is the one in which we must act in order to break out of the perceptual prison of those who study war as part of the human condition. Consciousness is also part of the human condition, and provides the fundamental reason why 'laws' can vanish under the weight of political will.

Further Reading

Becker, J. J. (1985). *The Great War and the French People*. Leamington Spa: Berg.

Berghahn, V. (1973). *Germany and the Approach of War in 1914*. London: MacMillan.

Ferro, M. (1973). *The Great War 1914–1918*. London: Routledge & Kegan Paul.

Fischer, F. (1967). *Germany's Aims in the First World War*. London: Chatto & Windus.

Joll, J. (1984), *The Origins of the First World War*. London: Longman.

McNeill, W. (1980). *The Pursuit of Power*. Oxford: Blackwell.

Mommsen, W. (1981). The topos of inevitable war in Germany in the decade before 1914'. In V. R. Berghahn & M. Kitchen (eds), *Germany in the Age of Total War*. London: Croom Helm.

Snyder, J. (1986). *The Ideology of the Offensive*. Ithaca, New York: Cornell University Press.

Stoessinger, J. G. (1980). *Why Nations Go to War*. London: MacMillan.

Taylor, A. J. P. (1958). *The Origins of the Second World War*. London: Allen & Unwin.

16

The political origins of international war: a multifactorial review[1]

J. DAVID SINGER

International and civil war, despite their *apparent* frequency, are relatively rare events. For example, it may be perfectly true that since the Napoleonic Wars there have been an *average* of six international and six civil wars per decade, but they have tended to be very unequally distributed over time as well as place (Small & Singer, 1982). Thus, while 'human nature' may make war possible, war is certainly neither universal nor ubiquitous.

This line of reasoning suggests, then, that our genetic heritage permits war, but is far from *sufficient*, and so it behooves us to discover – with all deliberate speed – which combinations of non-genetic factors can combine with our *capacity* for war to permit its occurrence. And, of course, our incentive for such research is not only to help understand those factors, conditions and behaviors that need to be avoided in order to minimize if not eliminate war, but also to reduce the extraordinary costs – material and otherwise – that are associated with the mobilization for war. That is, if it is still widely believed by political élites and ordinary people around the world that preparation for war is the most effective method of war prevention, we need be concerned with these costs as well as those that ensue from war itself.

Despite some pessimism about the effectiveness of scientific data in influencing national policies, there are several good reasons for encouraging rigorous and systematic research into the correlates and etiology of war. One of these is merely to demonstrate our scepticism about the conventional wisdom of our respective nations: some governments may want the population to continue to believe their statements or assump-

[1] A more fully documented version of the major portion of this paper was published under the title of 'Accounting for international war' in the Annual Review of Sociology, 1980.

tions about the virtues of their policies, the evils of others' policies and the need for military force to protect the nations' virtues and interests. A second is that in the early stages of a new research area, we make many more discoveries of a debunking sort than of a confirming sort, and it is quite important to demonstrate with reproducible evidence that the folklore is often incorrect. History seldom 'teaches' what the national security élites tell us it teaches. A third important reason is to encourage the systematic evaluation of alternative views of the global security problem, and if this can be done in the context of hard evidence and close reasoning, so much the better. A fourth, perhaps the most telling argument, is also the most obvious one: if we can indeed discover which policies under which sets of conditions have most regularly led the nations toward or away from war, and the evidence is fairly compelling, it might actually lead to more adaptive national policies. However, given the general ignorance of such factors, as well as the awesome host of factors that drive nations into destructive (and self-destructive) behavior, the resistance to such an approach is likely to be great at the outset.

Bearing these serious limitations in mind, while also recognizing the methodological, conceptual, theoretical and institutional difficulties of launching a new and politically questionable field of scientific research, let us turn to an overview of what has been learned so far. In so doing, it is essential to note that, by strict scientific standards, the evidence must be viewed as highly tentative. There has been little replication, there are frequent anomalies and inconsistencies in the findings, the validation of our key indicators has barely begun, and on top of all this, our theoretical models rest on insecure foundations. And yet, they constitute a basic step towards the differentiated and critical analysis of the problem that must replace 'mono-causal' approaches.

Typologies and taxonomies

Given this brief introduction to the emerging interdisciplinary field of war–peace studies, where do we stand now? What kinds of knowledge have we acquired and how adequately has it been codified into a theoretically coherent form? What are the dominant orientations, and which are the more promising ones? In order to address these questions, we need some sort of organizing framework, within which we can differentiate the several theoretical orientations as well as summarize our knowledge to date. The variety of frameworks is, of course, quite large, and people gravitate to one or another of them for all sorts of reasons, scientific or otherwise: nationality, social class, age, gender, education, personality, academic identity, foundation and governmental fashions, and even (!) prior research experience.

A familiar typology

One of the more familiar ways of classifying – and often selecting from among – the possible approaches is to divide them into the standard categories of technological, economic, geographic, political, demographic, ideological, psychological, etc. As suggested, the advocates of any of these approaches often arrive at their position as a result of disciplinary affiliation, reflecting the oft-implicit premise that the academic profession that one selects reflects an illusion that *its* major explanatory variables will account for virtually all sorts of social outcomes. Choosing such a typology and one of its approaches may, however, reflect the conviction that one best studies those phenomena that one knows best, and that such partial models and their findings must eventually be integrated into those generated within the other social sciences.

But regardless of the motives and assumptions, we end up with a typology that helps to perpetuate and legitimize these parochial orientations and to encourage the appearance, disappearance and reappearance of those 'theories' that are all too typical of the no-growth, non-cumulative disciplines. Thus, we find heavy, if not sole reliance on such factors as power discrepancies, surplus capital, business cycles, demographic pressures, resource needs, national moods and so forth. These at least have the virtue of resting upon variables that show *some variation* across time and place. But other putative models of a single-variable sort fail even to recognize that a phenomenon as irregularly (and infrequently) distributed across time and space as war cannot be explained on the basis of relatively *invariant* phenomena. Thus, it is difficult to take seriously such putative explanations as the human drive for power, or the instinct of aggressiveness.

In addition to the facts that most of the above orientations are rarely examined empirically, are seldom related to other explanations, ignore the multidimensional complexities of the war-inducing process, show insufficient variation in their 'explanatory variables', and are overly responsive to current events, political pressures and funding agency fads, they usually suffer from another fatal flaw. That is, they tend to overlook the critical distinction between international conflict and international war. While *conflicts* can arise out of an impressive range of social incompatibilities, the processes that lead to so frequent an event as conflict are not necessarily those that lead to so *infrequent* an event as *war*. In a global system that is so poorly integrated in the structural or cultural sense, the relatively high frequency of serious international conflicts and military confrontations (about 300 involving the major powers alone since the Congress of Vienna) need not surprise us, whereas the relatively low frequency of war (fewer than 30 among those same powers in those 160 years or so) surely calls forth our curiosity.

Then there is the distinction between necessary conditions and sufficient ones. Many conditions are necessary for modern war: the fact that humans can behave aggressively and that many of them do seek power or territory under the proper stimuli; the availability of transport and weapons technology; centralized decision authority; some sort of credible justification, and so forth. While these may well be *necessary* conditions, it is far from clear that any one of them is *sufficient* to carry a conflict across the threshold to war. We need a typology that recognizes the qualitative differences among competition, rivalry and dispute on the one hand, and military combat on the other, and that also recognizes the complex interplay of necessary and/or sufficient conditions reflecting a fair range of material, structural, cultural and behavioral phenomena. In addition, it should aid in integrating what we *have* discovered and hope to discover, and if possible, illuminate the research path that links the two. A discrete checklist will not suffice; an integrated, but multi-theoretical, framework is essential. A scheme has been derived from the general systems literature that might possibly help us to organize what we know, and think we know: to help stimulate the most appropriate next steps it has been laid out in considerable detail elsewhere, along with an elaborate epistemological rationale (Singer, 1971, 1980), but its bare outlines can be summarized here.

There is an almost limitless range of levels of 'social aggregation' at which the causes of war question might be addressed, but for the purposes of the discussion at hand, we will address only four of them: the international system (defined globally, regionally or functionally, depending on the work being summarized); the inter-nation dyad; the single nation; and the decision-making agency level.

'Social entities' from the individual up to the global system may be described in terms of three sets of attributes. In order of tangibility and ease of operational measurement, these are: material, structural and cultural. *Material* attributes may be divided into three subsets, of minimal relevance to the individual human, but highly germane to any other entity: geographic, demographic and technological.

By the *structural* attributes of a social system we mean the institutional and organizational phenomena by which economists, sociologists, and political scientists usually describe it. Structural properties may be of a relatively formal nature, embracing the types and powers of institutions that deal with legislative, administrative, judicial, banking, commercial, welfare, informational and similar activities in the entity or system. But they also embrace such informal attributes as access to and influence over decision-making processes; the number and configuration of political parties and pressure groups; the number and distribution of religious, linguistic and ethnic aggregations; the social and geographical mobility of the population; the extent and configurations of pluralistic cross-cutting bonds; and the nature and stability of the resulting coalitions.

By the *cultural* attributes of a social entity, we mean solely the distribution of the psychological properties of the individuals who comprise that particular system; this definition explicitly excludes, as unnecessary and scientifically misleading, any so-called emergent or organic properties of a psycho-cultural nature. The cultural dimension embraces the distribution of personality types, attitudes and opinions, and extends to the way in which all three of these psychological attributes relate to the way things are, should be and will be; these can be thought of, respectively, as perceptions, preferences and predictions. We can, in turn, treat the distribution of personalities as a national (or any other system's) character, that of attitudes as ideology, and that of opinion as cultural climate.

These dimensions should permit us to describe a nation or any social entity, compare it to itself at different points in time, and compare it to other entities at the same or different levels of analysis at the same or a different point in time.

While the distinction between behavior and interaction is self-evident, one of the inadequacies of the English and related languages and the concepts conveyed by its vocabulary is that the word '*relationship*' carries so wide a variety of meanings, but it is used here in only two of its senses: that of comparison and that of connectedness. Thus, when we say that the political integration of system X is very high in relation to that of system Y, we are comparing the two systems on a particular set of attributes. The confusion, as well as the distinctions, between these two meanings is well illustrated by the notion of 'social distance'. Two systems may be close or distant in terms of their position on a given *attribute* scale, such as size, ethnic homogeneity, or structural complexity. They may also be close or distant in terms of their *interdependence* or friendship, and there is very little reason to expect that entities that are close in the sense of similarity will always be close in the sense of connectedness or interdependence as well, despite the frequency of this assumption (implicit or otherwise) in the social sciences.

Measuring the incidence of international war

Before using this scheme to examine the evidence as to which factors may be related to fluctuations in the incidence of war, a brief digression is in order. Surprising as it may seem, scholars speculated on the causes of war for centuries before trying to ascertain its empirical distribution, and even today many of the theoretical disagreements stem partly from the failure to identify the population of cases, and from differing conceptions and definitions of the outcome variable. To rectify this situation, and following on the work of Sorokin, Richardson and Wright, the Correlates of War project at Michigan assembled what seems to be the total population of international (inter-state, imperial and colonial) and civil wars involving one or more sovereign states in any part of the world since the Congress of Vienna, and presented the coding rules, resulting data, and summary statistics (Singer & Small, 1972; Small & Singer, 1982).

In that handbook, we differentiate among four basic indicators of the incidence of war. The first is that of frequency, and is measured by the onset of sustained military hostilities between official armed forces of two or more sovereign national states culminating in at least 1000 battle deaths. The second is severity, and is measured by the number of battle deaths resulting from a qualifying war; the third is magnitude, measured in terms of nation-months of war; and the fourth is that of intensity, measured in battle deaths per nation-month or per capita. While frequency is measured only in the context of a given time and space domain, the other three indicators of the incidence of war can be applied to a given war as well as to all of the qualifying wars that occur in a specified time and space domain. In most, but not all, of the studies summarized here, the Correlates of War data base and indicators have been used, and unless the choice of indicator makes an appreciable difference, the results will be reported in terms of the simplest indicator of war: its frequency.

Yet another consideration when seeking to account for fluctuations in a given phenomenon is the extent to which those fluctuations show a discernible regularity across time or space. While I would not accept the proposition that goodness of fit between a given distribution and one or another statistical model permits a legitimate inference as to the *processes* that culminate in that distribution, such regularities can certainly be suggestive. Perhaps the most suggestive fit would be that of periodicity, on the assumption that a strong cyclical pattern might imply a degree of inexorability in international warfare. And, given the frequency with which cyclical patterns have been asserted, it may be worth reporting the results of a fairly systematic search for them. Simply put, we have found only the weakest trace of periodicity in the incidence of international warfare over these 160 years, with a barely discernible cycle of around 20 years. More important, when we look at the war experiences of the more war-prone nations *one at a time*, virtually every conceivable technique fails to produce evidence for any kind of regularity. Thus, it is perfectly true that there are peaks and troughs in the time plots of war at both the national and the systemic levels, but the time intervals between those peaks and/or valleys are of sufficiently random length to support the conclusions of other authors that periodicity is absent.

In sum, if war does not appear and reappear in a regular cyclical fashion, it is unlikely that it results to any important extent from any other single-factor cycle, be it commercial, agricultural, climatic or demographic. Rather, if there are indeed cyclical phenomena at work, there must be *several* of them involved in the process, with their concatenations falling at relatively irregular intervals. While there seems to be no concentrated research effort in this direction at the moment, it certainly appears to be worth pursuing further, perhaps when more is known regarding the time–space distributions of some of the more promising explanatory variables. Let us, then, attempt a brief and admittedly selective survey of these latter as they impinge on the incidence of international war.

Findings to date on the incidence of war

To reiterate a point which has already been noted, there is no substitute for a good theoretical model when laying out an empirical investigation, but when we are not nearly far enough along even to specify the key variables of such a model, we have little choice but to work within a multi-theoretical framework and get on with our more inductive efforts. The problem now is one of identifying the results of the inductive work to date, some (but not all) of which has been informed by the proposed framework, and to see how far it permits us to summarize and synthesize the diverse findings. We will work with three of the levels of aggregation noted earlier, and within each, look at the extent to which the several classes of variables seem to be accounting historically for the incidence of war. And, reflecting my strong suspicion that variables of a systemic and dyadic sort will turn out to be more powerful than those at the national and the decision-making levels of aggregation, let me deal with the findings at these levels in that order.

Systemic conditions

A fair fraction of research on the systemic conditions associated with war has emanated so far only from the Michigan Correlates of War project. Further, of the three types of systemic conditions– material, structural and cultural–most of the reproducible evidence to date reflects the structural dimension (Sullivan, 1976). Unfortunately, there is little systemic work on such *material* attributes of the system as weapons technology, industrial developments, resource limits, climate, or demographic patterns. Similarly, except for some preliminary analyses, little effort has been invested in the search for systemic connections between *cultural* conditions and the incidence of war. On the other hand, some scientific researchers have been as assiduous in their examination of structural correlates of war as their traditional colleagues, and it is to that literature that we now turn.

Perhaps the most plausible of the system's structural attributes in the war–peace context is that of the configurations generated by alliance bonds, with those generated by distributions of power following closely behind. Looking first at the structural characteristic known as bipolarity, we usually have in mind the extent to which the nations in a given geographical region, or in the major power subset (a functional 'region'), or worldwide, are clustered into two clearly opposed coalitions. While there are several definitions of bipolarity and rather diverse operational indicators, it generally implies the degree of conformity to an 'ideal' condition in which all of the nations are – via military alliance – in one or another of two equally powerful coalitions with *no* alliance bonds *between* the polar blocs, and *full* bondedness *within* each. While such a set of coalitions has never obtained, the reasoning is that even an approximation would so severely hamper the conflict-resolving efficacy of the pluralistic, cross-cutting multiple balancing mechanisms of the system as to make conflict escalation and war much more likely. But like all too many theoretical hunches in the world politics field, there is an equally plausible counter-argument: that so clearly bipolar a system structure would eliminate all ambiguity as to who is on whose side, or as to the possibility of military victory, that war would just never be considered. Rather, according to this orientation, war occurs when there is ambiguity, either because behaviour becomes less predictable and governments stumble into war, or because governments have a 'drive toward certainty', and war helps to clarify the picture.

In any event, the research findings to date only partially resolve the theoretical disagreements. In one of the first systematic analyses, Singer & Small (1968) found that the relationship between their indicators of polarity and war differed in the nineteenth and twentieth centuries. In the earlier epoch, higher polarity levels tended to be followed by *lower* levels of international war, but in the period since 1900, fluctuations in the incidence of war were *positively* associated with the fluctuations in polarity. In a follow-up, Wallace (1973) used a somewhat different indicator of polarity and found a curvilinear association, with war levels generally associated with very high bipolarity scores or very low ones, suggesting – for the entire 1816–1965 period – that there may be an optimal level of three or four relatively discernible alliance clusters, with war levels quite low when those intermediate conditions obtain. In a third analysis of the question, Bueno de Mesquita (1978) found that fluctuations in war were accounted for less by the *level* of bipolarity in the system than by the directions and rate of change in the alliance configurations that might produce such bipolarity. That is, the amount of war in the system since the Congress of Vienna tended to rise when the 'tightness' of alliance clusters was on the increase.

As might be expected, there are several interesting extensions of the bipolarity-war hypothesis. The simplest is that since alliances involving one set of major powers will generally be in opposition to other majors, the greater the percentage of major powers in alliance, the greater the bipolarity of the system as a whole, and thus the greater (or lesser) will be the incidence of war in the ensuing years. Once again, the empirical findings are mixed; we found that the alliance aggregation indicator, which was indeed highly correlated with the bipolarity index, predicted positively the incidence of war in the 1900–45 period, but negatively in the 1816–99 period.

A more complex version of this systemic hypothesis can be interpreted in two ways. One is that the concentration of military and industrial capabilities tends to follow the concentration of nations (especially the majors), and that the concentration of these *capabilities* should have the same effect on the incidence of war as the concentration of the *nations* into a small number of tight polar groupings. While this isomorphism does occur from time to time, it is not constant enough to make such an interpretation very compelling. Thus, one takes a more generalized view and treats both types of concentration as producing a high degree of clarity in the system's structure. And to the extent that the systemic environment is unambiguous as to (*a*) who will fight on which side if a conflict escalates to war, and (*b*) which side will probably win that war, the decision makers are thought to be less likely to either entertain the war option or to merely stumble into war.

Regardless of the theoretical interpretation, the empirical investigations lead once more to inconsistent results. And, as before, the major anomaly is the inter-century one. In the nineteenth century, high concentration of capabilities in the hands of a very few powers makes for increases in the incidence of war, while more equal distributions are associated with *low* levels of war. But in the period 1900–65, high concentrations lead to low levels of warfare and low concentrations are associated with higher levels of war. In a follow-up study, Champion & Stoll (1979) went a step further and (a) introduced an important control variable, classifying each major power as either 'satisfied' or 'dissatisfied' on the eve of each war; and (b) calculated the concentrations in terms of coalitions, rather than concentrations in terms of the nations separately and individually. These modifications appreciably enhanced the post-dictive power of the capability distribution model across the full time period, suggesting that if the blocs have indeed been accurately identified and the powers accurately classified on the 'satisfaction' dimension, this systemic factor may be of considerable importance. But as the authors remind us, the validity of these additional indicators remains to be more fully demonstrated.

Another factor – related to the others in the sense that it taps the structural clarity dimension – is that of status inconsistency, aggregated to the systemic level. Two of the earlier investigations found moderately clear associations between the incidence of international war and the extent to which the rank scores of the nations on the material capability and diplomatic importance dimensions were inconsistent with one another. That is, the more similar the systemic pecking orders on power and on prestige, the less war-prone the system was in the years following.

Dyadic conditions

While there is a clear conceptual difference between the structural characteristics of a system and the relationships among the component units of the system, it is worth reiterating their empirical and conceptual connections. Most systemic properties rest upon, and can be inferred from, the links and bonds among the components; while some scholars have sought their indicators of system structure in the triad, most of the data-based work has been restricted to the more manageable 'two-body problem', to borrow from the vocabulary of physics. Following the distinction noted earlier, we will treat dyadic conditions of a relational sort first, and then turn to those of a comparative sort, resting not on the links, but on the similarities, between nations.

As to the former, we again find the familiar emphasis upon alliance bonds, followed by the bonds created via membership in discernible diplomatic clusters, trading blocs, and international organizations. First, in general, nations with formal alliance bonds experience a significantly higher frequency of war than do those without them: on the other hand, nations that were allied with one another had a very low probability of going to war against one another. Looking at another type of bond, war opponents tended to decrease their shared intergovernmental organization (IGO) memberships in the 5-year period preceding the onset of war, and Singer & Wallace (1970) found that while most IGOs were founded after the termination of war, there was virtually no relationship between the number of them in the system and the amount of interstate war experienced in the *subsequent* 5 years. More surprisingly, on another dimension, serious disputes between major trading partners were *more* likely to escalate to war than were disputes between states that did not trade heavily with each other.

Shifting from the role of dyadic bonds and associations in accounting for the incidence of war to that of similarities and differences, the ubiquitous dimension of power again captures most of our attention. The theoretical argument is rather direct: even though sub-military conflict seems to be no more likely between nations of very unequal strength than between those of approximate parity, this factor should become more critical as the war threshold is approached. The familiar dictum is that the weaker *dare* not fight and the stronger *need* not; the corollary is that one purpose of war is to ascertain which party is stronger when any doubt does exist.

While the evidence on this question is relatively consistent, the final word is hardly in. For example, equality in population reduced the likelihood of dyadic war, but that equality in geographical size or industrial base did not, in the 1816–1965 period. When geographical contiguity was controlled, however, nations that were approximately equal in material capabilities were significantly more likely to carry their disputes to war than those of discernible *dis*parity. Further, serious disputes were more likely to escalate to war if the weak side was the initiator; disputes initiated by the stronger side were less likely to end in war.

A second emphasis in this literature is more diffuse, but worth noting briefly, given the theoretical pervasiveness of its assumptions. That is, the more similar two nations are in terms of certain political or cultural attributes, the more friendly their relationship might be expected to be, and the more friendly they are, the less frequently or severely might they be expected to wage war against one another. Richardson (1960*a*) found little historical evidence to support the classical view. For the period 1820–1949 and using his population of about 300 wars and military disputes, he found that neither a common language nor a common religion had a depressing effect on the incidence of dyadic war (pp. 230–31). To the contrary, as he himself demonstrated (pp. 285–86) and as others have confirmed, geographical contiguity is the confounding variable. That is, since geographical neighbors are not only more likely to be culturally similar but also to have more sources of conflict and to be more accessible to one another's armies, it follows that such similarities should actually be related to dyadic war in the *positive* direction.

These findings lead, in turn, to another of the more interesting paradoxes in research to date. Reference is to the effect of common boundaries, with the reasonable hypothesis that the greater the number of immediate neighbors a nation has, the more frequently it will be drawn into warfare against one or another of them. While Richardson's data tend to support this hypothesis, the findings of Starr & Most (1978) do not. Rather, they find an inverse relationship between a nation's war proneness and the number of immediate neighbors. Nor should these results be surprising, when we consider that the number of direct neighbors is physically a function of a nation's geographic size and that of its neighbors. The longer its boundaries, the greater the number of nations that can border on it, and the smaller these latter are, the more numerous they can be. From this, it follows that the greater will be the discrepancy between its size (and strength, all else being equal) and theirs, and given the finding that war is more likely between equals, it again follows that the frequency of war *should* be lower.

National conditions

In an earlier section, I indicated that national attributes seemed less crucial in accounting for war than either systemic or dyadic conditions, and before summarizing the evidence to date, let me expand on that assertion. Briefly put, the exigencies of survival in an international system of such inadequate organization and with so pervasively dysfunctional a culture require relatively uniform responses. That is, for a national entity to adapt to and survive in such an environment, it must achieve a fair degree of political mobilization, military preparedness, and political centralization. Despite great differences in tradition and culture, or great apparent differences in political regime and economic arrangement, these merely mute the essential domestic similarity of national states, regardless of their size, strength, level of economic development, etc.

To what extent does the empirical evidence to date support this alleged lack of variation in the national attributes that might be associated with the incidence of war? On the one hand, certain of these attributes do seem to be related to the frequency and/or severity of national war experiences, with overall military–industrial capabilities the most potent predictor. In a systematic examination of the relationship between a six-dimensional index of such capabilities and war proneness since the Congress of Vienna, Bremer (1979) found a very strong positive correlation between the strength of nations (including industrial development and military preparedness) and their tendency to go to war. Using a more restricted indicator, reflecting the size of the military establishment, and using more general indicators of foreign conflict, several studies further confirm this positive association with national strength.

On the other hand, if we turn to more complex models that might link national characteristics to war proneness, the findings are considerably more ambiguous. Applying the sociological concept of status inconsistency to nations whose material capabilities are high and whose attributed diplomatic status scores are low (or vice versa), two studies have failed to find any consistent pattern. Another domestic characteristic that has often been thought of as contributing to national war proneness is that of domestic instability. Looking at regime type alone and examining the full 1816–1965 period, we found that autocratic and democratic regimes were equally likely both to initiate wars and to become embroiled in them.

Picking up the geographical variables summarized earlier in the dyadic context, Richardson (1960a) found a positive relationship between the number of neighbors a nation had and the frequency of its wars from 1820 to 1945, and relatively similar patterns have been found for other comparable spatial–temporal domains. Finally, we examined the effect of a nation's mean distance from all other sovereign members of the system, and there, too, found a positive relationship; the more centrally located they were, the more war they experienced.

The tentative inference from this admittedly limited set of studies is that such basic geo-strategic factors as location and strength seem to be of importance, but that domestic factors of a less material sort would appear to be rather negligible in accounting for the war proneness of individual nations.

Behavioral and interactional patterns

To this juncture, our focus has been on the extent to which the frequency and magnitude of war might be accounted for historically by the ecological variables: fluctuations in systemic, dyadic or national conditions. Following the checklist implied in the taxonomy, this leaves untouched the question of behavior itself: to what extent can we account for war by the actions and interactions of the nations?

There is a modest body of empirical work in which behavior patterns serve as predictors, and once again Richardson offers a convenient point of departure. Perhaps his most important contribution is found in the posthumous *Arms and Insecurity* (1960*b*), where he derives and then puts to the test a simple differential equation designed to capture the essence of an interactive arms race. *Vis-à-vis* the arms expenditure patterns preceding the two World Wars, the model offers a fairly good fit, and generally supports the notion that each protagonist's annual increase will be a function of the other's absolute expenditure in the previous year, controlling for a fatigue factor and an exacerbation factor.

These analyses have stimulated the development of a rich and diverse array of follow-up models, a fair number of which have been tested against various nineteenth and twentieth century arms interaction processes. While it would require a major review article to summarize and interpret this body of research, two general conclusions seem justified. One is that we have not yet been able to separate out the effects of the domestic and foreign stimuli at the various stages of arms races, and the other is that we have yet to differentiate between the profiles or 'signatures' of those that have ended in war and those that have not. But two important findings have resulted from Wallace's work in this area. In an earlier study, he not only found that status inconsistency levels in the international system predicted to system-wide increases in military expenditures, but also that these increases predicted in turn to the incidence of war. In a later analysis (1979), Wallace concluded that if two nations found themselves in a military confrontation, their likelihood of crossing the war threshold was considerably lower if they were not in an arms race with one another, but much higher if they were.

An even more elusive problem arises when we shift from military expenditures and arms acquisition to less easily observed behavior such as the diplomatic moves and counter-moves associated with the escalation of conflict. But a satisfactory coding and scaling scheme has been developed and in some preliminary analyses, Leng (1979) found that the use of threats had a higher probability of ending in war when met with a defiant counter-threat, whereas there is no association between the mere frequency of threats and the onset of war. Not surprisingly, and despite the attention of diplomatic historians and traditional political scientists, the systematic examination of the relationship between behavioral phenomena and war has lagged discernibly behind that of the other three sets of factors discussed earlier. While the explanation lies partially in the observation and measurement problem, this lag also reflects the reasonable idea that the more we first discover regarding the effects of the ecological conditions, the greater the theoretical mileage we will obtain from subsequent analyses of the behavioral and interactional phenomena.

In this connection, a comment on the ecological discontinuities between the pre-nuclear and the nuclear eras might be useful. My colleagues and I are finding a pattern that is both promising and alarming. First, the 'bad news': (*a*) there is a slight increase in the frequency with which major powers become involved in militarized confrontations since 1945; and (*b*) they initiate these confrontations and respond to one another's moves in very much the same way that they did in the 1816–1945 period, showing a strong drive for 'escalation dominance'. Further, they build, deploy and talk about nuclear weapons as if they were indeed as usable as any 'conventional' weapon of mass destruction. But the 'good news' is by no means negligible: whereas more than 10 per cent of all major power confrontations escalated to all-out war from the Congress of Vienna until World War II, not a single one has so erupted since the advent of the nuclear age.

While some would interpret the latter as a sign of assurance and hope, such would seem premature in light of the first set of patterns. Given the continued tendency to brandish these new weapons as if they were merely 'big cannons', the reluctance to reduce and eliminate them, and the fast-changing technological environment, my view is that World War III is hardly less likely than were World Wars I and II.

Conclusion

While this summary hardly suggests that we are well on the way to understanding the causes of war and conditions of peace (to use Wright's phrase), some modest optimism would not be completely out of place. That is, despite the limited amount of research and the relative lack of convergence in our findings, there are two grounds for believing that we are no longer moving in non-cumulative circles. First, we are finally seeing systemic research that rests on reproducible evidence, and after several centuries of pre-operational speculation, this is to be applauded. Second, and most relevant for the volume at hand, is the long-needed involvement of those researchers who appreciate the meaning of 'human nature'. Historians, political scientists and econo- mists have for centuries been invoking this concept in a manner that is more satanic than sophisticated, and more theological than theoretical. And the consequences of such ignorance have been – and remain – a menace not only to competent research but also to adaptive policies. By illuminating the link and the interface between individuals and their social systems, 'aggression research' may well clarify that elusive process by which individual human infants are so often molded and mobilized into collectivities of lethal warriors. And third, despite the diversity of theoretical orientations, there seems to be a growing awareness that the resulting paradigms and investigations need not be incompatible.

On the one hand, my sense is that researchers in the war–peace sector are increasingly aware of the difference between explaining the *high* incidence of recurrent conflict and rivalry of a non-military sort, and explaining the low – but destructive – incidence of war. The range of models intended to address the former question is, admittedly, impress- ive in its diversity. But that, in my judgement, is a less critical problem; conflict is ubiquitous in all social systems, and the problem is not to prevent it, but to reduce the frequency with which it becomes socially destructive. Thus, when we shift from conflict in general to war in particular, we find a diminished and more manageable array of theoreti- cal models. And, on the other hand, a close scrutiny of these models shows a remarkable overlap, if not convergence. To illustrate, let me mention some of those found in the literature most frequently: realpoli- tik, arms race, power transition, economic development, and imperial- ism. There are indeed those who insist that such models are not only inconsistent with one another logically and empirically, but also not even subject to comparative analysis and testing. There are even those – especially among devotees of the 'imperialism' model in its several forms

– who insist on a different epistemology and assert that the western scientific method is inappropriate to the examination of rival models and hypotheses. But my suspicion is that their numbers and influence are on the wane, and that a compelling theoretical convergence could encourage a more open-minded examination of the empirical evidence.

In order to move toward such a theoretical convergence, two prior steps would seem necessary. First, we need to identify the extent to which these competing models rest upon similar explanatory variables. Given the Babel-like discourse of theorizers and practitioners, this might appear to be impossible. But as systemic empirical work goes forward, one inexorable consequence is the translation of ideologically loaded verbalization into operationally defined variables, and the evidence to date is that devotees of rather diverse approaches can agree on (at least) the face validity of these indicators. Moreover, it looks as if a rather wide range of putatively explanatory concepts can be validly translated into identical indicators. Thus, 'defining our terms' and operationalizing our variables may well reveal a considerable degree of convergence among theoretical models that have often been thought of as hopelessly divergent and incomparable.

Second, we need to recognize that at rock bottom the most important difference among the contending causes of war models is that of the foreign policy decision process. That is, each model assumes – often implicitly – a different class of decision makers in power, and each postulates a different set of decision rules. Thus, one strategy might be to conceive of the policy-making process in terms of interest aggregation, with the decision makers seen as trying to respond to and balance a complex array of international incentives and constraints *vis-à-vis* an equally complex array of domestic interests, including their own. Once those assumptions and postulations have been teased out of the verbal models, our rather highly general paradigm might then be converted into operational versions of the several more specific models.

They cannot, however, be put to the empirical/historical test directly, and this is why the research described in the body of this paper can be so valuable. It is one thing to observe and then classify the backgrounds and interest group affiliations of foreign policy élites in a wide range of nations across an appreciable span of time; while this research task is far from complete, it is clearly manageable. But the decision rules that they employ, and the utilities that they assign to various outcomes in the economic, diplomatic and military sectors will probably remain forever beyond the range of direct observation. Thus, we have little choice but to rely on careful inference, and the most solid basis for inferring their decision criteria will be found in the sort of evidence discussed above. *Under which systemic, dyadic and national conditions do which foreign policy élites respond to which behavioural stimuli of which other nations in which specific fashions?* The more fully we can answer these empirical questions, the more reliably can we infer the decision rules that are at work. And the more we can check these inferences against those that are embedded in the rival models, the closer we can get to their confirmation or disconfirmation.

This brief overview leaves out a great deal of important detail, just as our survey of the evidence to date is far from complete. But these pages should serve to remind us that the causes-of-war question is not merely a researchable one, soluble in principle. They also serve to remind us that the task is finally under way, and if the scientific talent and the necessary support can be mobilized, we may yet be in time to put an end to one of the most destructive and dysfunctional activities known to human history.

References and further reading

Bremer, S. (1979). National capabilities and war proneness. In J. D. Singer (ed.), *Correlates of War II: testing some realpolitik models*. New York: Free Press.

Bueno de Mesquita, B. (1978). Systemic polarization and the occurrence and duration of war. *Journal of Conflict Resolution*, 22/2, 241–67.

Champion, M. & Stoll, R. (1979). Capability concentration, alliance bonding and conflict among the major powers. In A. N. Sabrosky (ed.), *Alliances and International Conflict*. Philadelphia: Foreign Policy Research Institute.

Leng, R. J. (1979). Influence strategies and interstate conflict. In J. D. Singer (ed.), *The Correlates of War II: testing some realpolitik models*. New York: Free Press.

Richardson, L. F. (1960a). *Statistics of Deadly Quarrels*. Pittsburgh: Boxwood.

Richardson, L. F. (1960b). *Arms and Insecurity*. Pittsburgh: Boxwood.

Singer, J. D. (1971). *A General Systems Taxonomy for Political Science*. Morristown, New Jersey: General Learning Press.

Singer, J. D. (1980). Accounting for international war. *Annual Review of Sociology*, 349–67.

Singer, J. D. & Small, M. (1968). Alliance aggregation and the onset of war 1815–1945. In J. D. Singer (ed.), *Quantitative International Politics: insights and evidence*. New York: Free Press.

Singer, J. D. & Small, M. (1972). *The Wages of War: a statistical handbook 1816–1965*. New York: John Wiley & Sons.

Singer, J. D. & Wallace, M. (1970). Intergovernmental organization and the preservation of peace, 1816–1964: some bivariate relationships. *International Organization*, 24, 520–47.

Small, M. & Singer, J. D. (1982). *Resort to Arms: international and civil war, 1816–1980*. Beverly Hills: Sage.

Starr, H. & Most, B. (1978). A return journey. *Journal of Conflict Resolution*, 22/3, 441–68.

Sullivan, M. (1976). *International Relations: theories and evidence.* Englewood Cliffs, New Jersey: Prentice-Hall.

Wallace, M. D. (1973). Alliance polarization, cross-cutting, and international war, 1815–1964. *Journal of Conflict Resolution*, **17**, 4.

Wallace, M. (1979). Arms races and escalation: some new evidence. *Journal of Conflict Resolution*, 23, 3–16.

Wright, Q. (1942). *A Study of War*, 2 vols. Chicago: University of Chicago. (Revised edition, 1965).

F. CONCLUSION

17

A multi-level approach to the problems of aggression and war

JO GROEBEL AND ROBERT A. HINDE

Aggression and war cannot be described in monocausal terms. As the chapters in this book show, studies from a variety of scientific disciplines demonstrate that individual aggression, group aggression and war can be moderated by biological, psychological, group and societal factors. No one of these factors is ever sufficient to explain the occurrence of aggression or war, and it is necessary to develop a multidimensional approach. Such an approach must come to terms both with the multiplicity of causes of simple aggressive acts and with the differences between the nature and determinants of aggression at different levels of complexity.

It is necessary to start with the recognition that human social behaviour involves a number of levels of social complexity (Chapter 1). At the individual level, biological and psychological factors interact with each other and with aspects of the current situation. Aggressive behaviour involves at least two individuals, and what occurs depends on both of them. If the two individuals are known to each other, what occurs will also be affected by their relationship, and will in its turn affect that relationship. And the course of that relationship will also be influenced by, and influence, the group in which it is embedded. And that group will interact with other groups within a larger society.

It is crucial to distinguish between these levels of social complexity in part because each successive level has properties that are simply not relevant to the level below. For example, the relationships within a group may be linearly, centrifocally or hierarchically arranged, considerations that have no relevance to individual relationships. The distinction is crucial also because different processes operate at each level: for instance, the dynamics of groups involves principles that are irrelevant and/or unnecessary for understanding the behaviour of individuals. These distinct sets of principles at each level provide as it were a second dimension, orthogonal to the distinction between levels of complexity, for an adequate approach to human aggression. These principles must concern the development of the potential for aggression, the elicitation of aggression, and the consequences of aggression on future aggressiveness, at each level.

But while it is important to distinguish between these levels of complexity, it is also important not to reify them. Groups, relationships, even individuals are to be seen as processes, maintained in equilibrium or changed through their mutual influences upon each other. These influences operate in large measure through the collective attitudes, beliefs and values of individuals, and through the institutions with their constituent roles within the society. This 'socio-cultural structure' influences the behaviour of individuals, but at the same time it is created by individuals and its characteristics are a consequence of the behaviour of individuals.

In the preceding chapters, we have seen how these generalities apply to the specific issues of aggressive behaviour at the successive levels of social complexity. *At the individual level*, biological and psychological factors interact with environmental influences in the development of a potential for aggressive behaviour. Except in some pathological cases, that potential, due in part to genetic factors (Chapter 5) and depending upon neural and hormonal influences (Chapter 6), is not necessarily realized in action (Chapter 4). It may be enhanced or diminished as a consequence of experience: here, childhood experience is of special importance (Chapters 7 and 9). But throughout life, individuals incorporate beliefs and values from, or adjust pre-existing beliefs and values under the influence of, the socio-cultural structure mediated by relationships with other individuals (e.g. Chapter 10).

Whether or not an individual behaves aggressively in any particular situation depends in part on the characteristics of that situation (Chapter 8). Many of the situations that elicit aggression can be described as frustrating, but whether aggression or some alternative behaviour appears depends in part on how the individual interprets the situation. If the frustration can be attributed to an identifiable source, such as a rival, an aggressive reaction is more probable. And the consequences of an aggressive response affect the probability of further aggression, though the effectiveness of particular consequences are influenced by the values and norms of the society. For instance, in some school situations, fighting has been seen as a usual way for adolescent boys to settle their disputes and, according to the norms of the society, violence between husband and wife may be condoned or prohibited. Again, in some cultures, societal values may require an aggressive coalition between two blood relatives against a third party deemed to have insulted one of them. The habit of aggression develops in interaction with this socio-cultural system and also contributes to the constitution of that system. For instance, if aggressive acts are seen to be successful, the values inherent in the society may become more favourable to them.

At the group level, the aggressive propensities of the individuals within the group affect the propensity of the group as a whole to behave aggressively. However – and this may be another way in which humans differ in degree from other species (Chapter 3) – the aggressiveness of a group is not simply the sum of that of its members: the issue is made much more complex by properties inherent in human groups (Chapter 11). The very fact of belonging to a group leads individuals to see their own group as favourably distinct from other groups, to exaggerate differences between their value systems (Tajfel, 1978), and thus to become more prone to aggressive acts against other groups. Furthermore, cohesion, centrality and the power distribution within the group may affect the probability of group aggression. Thus, an hierarchically structured group led by an individual with strong aggressive propensities will be more likely to show aggression. This is not merely a matter of coercion by the leader. If aggression is valued in the group, individuals can gain in status by being seen to behave aggressively. And reciprocal influences also operate: an aggressive group is likely to prefer an aggressive leader.

Thus, given individuals with a propensity for aggression, and given certain characteristics of group structure, a climate is created which can lead to aggressive activity if external and internal elicitors are present. A perceived threat against the interest or values of the group could serve as an external instigator. But an external instigator is not essential for intergroup aggression: members of a group may seek a target in order to show off their aggressive prowess to fellow group members. Indeed, a decrease in group cohesion can act as an internal elicitor: Groebel & Feger (1982) have shown that, in such circumstances, German terrorist groups were especially likely to show aggression.

As with individuals, if aggressive activity against another group turns out to be a successful, group values favouring aggression will be reinforced and aggression will become more likely. Furthermore, a successful aggressive group will become more attractive to individuals with aggressive predilections, who may then join it.

Thus, the history of the group and the way in which it defines itself, as well as its values and perceived needs, its structure, and the characteristics of the individuals within it all affect its potential for aggression in particular circumstances.

A societal system can facilitate or impede the occurrence of aggressive groups within it. One issue, of course, concerns the distribution of material and social resources: a distribution perceived as just is an important though not a sufficient condition for decreasing the likelihood of aggression. But most societies are heterogeneous, some constituent groups having more power than others and thus contributing more to the predominant value system. Powerful groups may impose values on individuals that run counter to the interests of those individuals. If the more powerful group or groups have an aggressive structure, it becomes more likely that the socio-cultural structure will embrace aggressive norms. Aggressive groups or individuals may play an important role here symbolically or by providing charismatic leaders. Whether or not the socio-cultural structure is conducive to war is of course affected also by other factors, especially historical and religious ones, as we shall see in a moment. Given the propensity to use aggressive means, a threat to the societal system, for instance either economically from international competition or by an increasing loss of intra-societal stability, may predispose the group in power to promote war against the source of the threat.

Turning then to the level of *international war*, although aggressive individuals and groups may play some role in its occurrence, the behaviour of individuals in modern war is for the most part influenced by their aggressive propensities only indirectly. With each level of social complexity, new factors enter, making it more difficult to make clear-cut causal statements about the origins of aggression (Chapters 15 and 16). This complexity may itself be an important cause of aggression. Some studies (e.g. Dörner *et al.*, 1983) have shown that even individuals, though normally unaggressive, may choose aggressive means to 'solve' a situation which they could not manage because of informational or situational overstrain. And the complexity of modern societal, economic and political situations in itself poses a danger, for it induces a tendency to simplify argumentation and create a subjective perception of manageability, even when it comes to war. Cutting the Gordian knot may be a metaphor for creativity, but it may also lead to destruction as a consequence of complexity.

Modern war is institutionalized, and although it involves aggressive acts against other nations, the behaviour of individuals is influenced by their aggressive propensities for the most part only very indirectly. As we have seen, the institution of war prescribes a variety of roles, each with its attendant rights and duties. Politicians, generals, soldiers, munition workers perform their allotted tasks, carrying out their duties with little contribution from their aggressive propensities. This is true even of the combatants, for whom cooperation and buddy-relationships, obedience and fear may be more important than aggression.

However, the institution of war must not be accepted as given. If we are to neutralize its power, we must understand its sources (Chapter 14). And these must surely be sought amongst the behavioural propensities of individuals. In large measure, these are mediated through the forces of history and religion that created the current socio-cultural structure with its institution of war. One is taught to honour those who died in past wars which created the present situation, or failed to create a better one. In honouring them, one must respect their cause and thus the institution of war that they served. The role of religion in maintaining the institution of war was apparent long before the Crusades and remains potent to the present day. Religion and ideology may be so strong that 'defending' them by attacking an enemy with another belief system is perceived as a personal goal.

But individual behavioural propensities enter more directly into the maintenance of the institution of war. Identification with the society's value system increases the individual's feeling of belonging and security. Propaganda, created by the individuals or groups in power, uses aggressive images and plays on the aggressive propensities of individuals (Chapter 13). It also plays on their fear of strangers, utilizing propensities to denigrate the outgroup to create an image of an enemy who is dangerous because almost non-human. Society in itself creates material rewards – medals, honours, promotion – that appeal to basic human propensities but at the same time stabilize the institution of war. Complex positive and negative reinforcement systems operate, individuals avoiding the destabilization of their values, isolation and punishment and gaining personal success and material advantages by supporting the institution of war. Those who stand back are disgraced.

Many modern institutions, and especially that of war, create and are created by an increasingly complex technology. Here lies another danger. Many incidents in recent decades demonstrate that the most sophisticated alarm and weapons systems are not infallible. The complex systems are susceptible to unpredictable errors which could lead to war even in the absence of human intent.

But in these facts lies hope. If aggression and war have been constituted by human action and depend on human propensities and human technology, then by human activity they can be modified and controlled (Chapters 9 and 10). Other destructive phenomena, believed for long to depend on natural laws, such as slavery, cannibalism and public executions, are now nearly extinct in most parts of the world: violence and war can be similarly reduced (Hugh Middleton, pers. comm.)

In the long run, it is the myths that are largely responsible for both individual aggression and war (Chapter 2 and 12). We may believe that our behaviour is mostly guided by rational principles. But despite our apparent enlightenment, many of our thoughts, decisions and behaviours are determined by rules other than those following mathematical logic. To organize and simplify our life and our environment, myths, that is non-verified assumptions and beliefs about cause–effect relationships, are created and maintained in our mental construction of the world.

There are many of those myths. The myth that aggression is a simple innate drive. That violence follows a natural law. That individual aggression, group violence, and war are completely analogous and follow the same rules. That wars are necessary parts of human life and follow a regular periodicity. That what has been true and sometimes successful in wars of the past has to be true and successful forever. There is the myth that our own values are the only valid ones. That complex systems are completely controllable. That our behaviour is mostly guided by rational thinking. That the world is just as it is. Some of these myths can readily be falsified by scientific study. Others have to be discussed in their specifics. By reducing the myths, by being self-critical about our own beliefs, by increasing communication with members of other groups and societies, learning about their views and establishing mutual confidence, we can contribute to a less aggressive world.

The abolition of violence and war demands a systematic analysis of their causes, a falsification of the myths surrounding them and especially an efficient search for non-violent alternatives to conflict resolution. As a start, at the individual level, we can examine practices of child-rearing and the causation of individual aggression; on the group level, we must establish increased communication between groups and an understanding of the social forces that operate; and on the societal level, we must seek for a policy where violence as political means is proscribed. All this cannot be done merely with a missionary's goodwill: rather it requires an approach involving a clear-headed approach on the several interacting levels of social complexity.

References and further reading

Dörner, D., Kreuzig, H. W., Reither, F. & Stäudel, T. (eds) (1983). *Lohausen. Vom Umgang mit Unbestimmtheit und Komplexität.* Bern: Huber.

Groebel, J. & Feger, H. (1982). Analyse von Strukturen terroristischer Gruppierungen. In *Analysen zum Terrorismus, Band 3: Gruppenprozesse*, pp. 393–432. Opladen: Westdeutscher Verlag.

Hinde, R. A. (1987). *Individuals, Relationships and Culture: links between ethology and the social sciences.* Cambridge: Cambridge University Press.

Miller, N. E. (1964). *Physiological and cultural determinants of behavior.* Washington, D.C.: Proceedings of the National Academy of Science.

Selg, H., Mees, U. & Berg. D. (1987). *Psychologie der Aggressivität.* Göttingen: Hogrefe.

Tajfel, H. (ed.) (1978). *Differentiation between Social Groups.* London: Academic Press.

Väyrynen, R. (1987). *The Quest for Peace.* London: Sage.

von Clausewitz, C. (1832–1834). *Vom Kriege.* (English translation: *On War.*) Berlin.

Name Index

Numbers in **bold type** indicate bibliographical references.

Subject Index